PNEUMATICS
AND HYDRAULICS

by Harry L. Stewart

A REVISION OF
FLUID POWER
by Harry L. Stewart

THEODORE AUDEL & CO.
a division of

HOWARD W. SAMS & CO., INC.
4300 West 62nd Street
Indianapolis, Indiana 46268

THIRD EDITION
Fourth Printing—1981

Copyright © 1966, 1968, and 1976 by Howard W. Sams & Co.,
Inc. Indianapolis, Indiana. Printed in the United States of
America.

International Standard Book Number: 0-672-23237-5
Library of Congress Catalog Card Number: 75-36658

Foreword

The purpose of this book is to aid in providing a better understanding of the fundamentals and principles of pneumatics (air) and hydraulics (oil), usually referred to as fluid power, and their related devices. A knowledge of pneumatics and hydraulics has become more important with the growing number of applications of pneumatic and hydraulic equipment in our expanding economy.

The applied principles and practical features of pneumatics and hydraulics are discussed in detail. The installation, operation, and maintenance of both hydraulic and pneumatic devices are covered thoroughly.

The information in this book has been compiled primarily to help engineering students, junior engineers, designers, process planners, installation and maintenance technicians, shop mechanics, tool men, shop foremen, industrial arts teachers, salesmen of pneumatic and hydraulic products and systems, and others who are interested in technical education and self-advancement. An appendix is included as a source of reference to related data.

The author is sincerely grateful to the many individuals and industrial organizations who have so generously provided illustrations and information.

HARRY L. STEWART

About the Author . . . .

Harry L. Stewart has been engaged in the Fluid Power industry for more than 30 years. He has served in research, product development, engineering sales, advertising, sales management and marketing. Mr. Stewart has lectured before many technical societies, plant operating groups, and educational groups in the United States and Canada, and has written more than 150 feature articles on fluid power, controls, and power-operated holding devices for technical magazines in the United States and Great Britain. Also, he is the author of *Hydraulic and Pneumatic Power for Production*, printed in English and Spanish and coauthor of five books published by Howard W. Sams & Co—*Fluid Power, ABC's of Hydraulic Circuits, ABC's of Pneumatic Circuits, Fluid Power Student's Workbook,* and *Fluid Power Instructor's Manual.*

Mr. Stewart is a past member of the Educational Committee of the *National Fluid Power Association* and has had a long association with the *Fluid Power Society* and the *SME.* He has traveled extensively in North America promoting fluid power and its uses in industry.

Contents

CHAPTER 21

Record of performance of components—repair parts availability—preventive maintenance.

CHAPTER 22

Control systems—differential sensing or error-detecting devices —types of servo systems—characteristics of servo systems.

APPENDIX

CHAPTER 1

Introduction to Basic Devices and Systems

Many different machines and processes use a fluid for developing a force to move or hold an object or for controlling an action. In automobiles, for example, hydraulic brakes are used to stop the car. In road construction and repair, another example, compressed air is used to operate chipping hammers. Many fluid units are employed in modern industry. Countless applications can be cited. Machines and processes are becoming more and more automated in order to meet competition and to reduce human error.

HYDRAULIC AND PNEUMATIC SYSTEMS

Generally speaking, a number of fluids can be used in devices and systems. The term *hydraulic* refers to a liquid. For example, the term *hydraulic turbine* can be used to designate a turbine involving water flow. In a hydraulic system, oil, water, or other liquid can be used. Besides a liquid, either a gas or a compressible fluid can be used. In actual practice, two fluids are most commonly used; there are oil and compressed air. Thus, when the word *fluid* is used in this book, oil and air are referred to primarily. A fluid system that uses oil is called a "hydraulic" system. A fluid system that uses compressed air is called a "pneumatic" system. Both types of fluid systems are treated in this book.

Hydraulic Systems

Fig. 1 illustrates a typical hydraulic system. Oil from a tank or reservoir flows through a tube or pipe into a pump. The pump

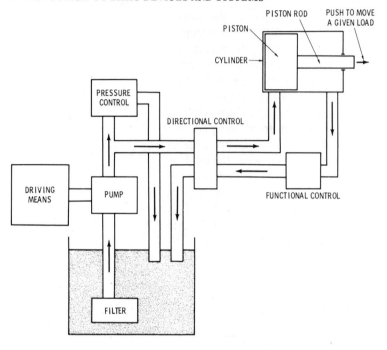

Fig. 1. A typical hydraulic system.

can be driven by an electric motor, air motor, gas turbine, or internal combustion engine. The pump increases the pressure of the oil; the oil pressure at the pump outlet may be 5 to 5000 or more pounds per square inch. High-pressure oil flows in a tube or piping through a control valve; this valve can be used to change the oil flow. A relief valve is used to protect the system; the valve can be set at a desired safe maximum pressure. If the oil pressure in the system begins to rise above the maximum safe pressure, the relief valve opens to relieve the pressure and to prevent damage to either the system or the surroundings. The oil that enters the cylinder acts on the piston; this pressure action over the area of the piston develops a force on the piston rod. The force of the piston rod can be used to move a load or device. Oil from the cylinder returns to the reservoir. As the oil passes through the filter, dirt and foreign matter are removed from the oil. Each separate unit, such as the pump, the valve, the cylinder, or the filter is called a "component" of the system.

There are certain advantages in the use of oil as the working fluid. Oil helps to lubricate the various sliding parts, such as piston elements, in the cylinder. Oil prevents rust and is readily available. For practical purposes, oil is a liquid that does not change its volume in the hydraulic system when the pressure is changed as the oil moves from one part of the system to another. If the oil fills the system completely, the movement of the piston can be controlled very closely by the oil flow.

For example, in the hydraulic system shown in Fig. 1, it can be assumed that the oil pressure at the inlet to the cylinder is 1500 pounds per square inch and that the area of the piston over which the oil pressure acts is 2 square inches. Thus, the force of the oil on the piston is (2 × 1500), or 3000 pounds. This indicates that a relatively large force can act on a load for a relatively small size of cylinder. This is one advantage of hydraulic devices.

Pneumatic Systems

Fig. 2 illustrates a pneumatic system that uses compressed air. Air from the atmosphere flows into the inlet of the air compressor. The air compressor increases the pressure of the air; at the compressor discharge outlet, the air pressure may be nearly 90 pounds per square inch greater than the atmospheric pressure (90 pounds per square inch "gauge"). The air compressor may be driven by either an electric motor or an internal combustion engine. On road construction, for example, a portable air compressor driven by an engine may be used. A relief valve at the compressor discharge is used to avoid dangerously high pressures. A filter in the system removes dirt from the air. A lubricator in the circuit adds some oil to the air passing through, and lubricates any sliding surfaces, such as the piston and cylinder surfaces. The compressed air acts on the piston to develop a force on the piston rod for moving a device or load. Fig. 2 illustrates a piston that moves in a straight line. An air cylinder or "air motor" can be arranged with a rotating member.

The use of compressed air has certain advantages. An air device presents no spark hazard in explosive atmospheres. An air device

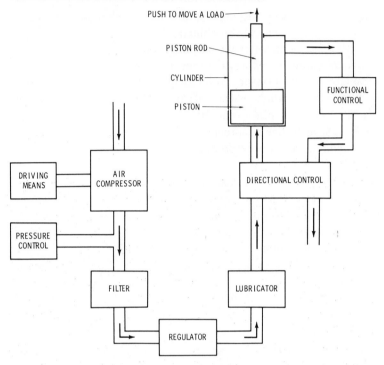

Fig. 2. A typical pneumatic system.

is used under wet conditions to avoid the electric shock hazard. For example, compressed-air power is the only type of power used in some mining operations. Air is readily available, because compressed air can be stored in a tank for instant use; a small compressor can be used to fill a storage tank for intermittent use, and return lines are not needed.

FLUID SYSTEMS

A large number of circuits or systems can be devised, and a large number of different components can be used. Essentially, there are three main or basic features in the usual simple fluid system: (1) an oil pump or an air compressor; (2) a device with either a piston or a rotating member driven by the fluid; and (3) piping and valve devices to control the flow of the fluid.

14

With these basic features of a simple system as a beginning, one can imagine various combinations. Several simple systems can be combined. One or more oil pumps can be used to actuate one or more cylinders. More than one valve can be used. In a plant, one large air compressor can be used to supply air for a large number of separate cylinders, each performing a different task.

The actuating piston can be given any movement that is required. Straight-line reciprocating motion is most often needed. Rotary motion can also be provided with various forms of hydraulic and air motors.

A fluid system has certain characteristics that are important in meeting certain service requirements. It is relatively easy to connect one component with another by tubing or piping. In some instances a flexible hose is used. A fluid can be used to cushion shocks. Many actions can be controlled by a simple manipulation of valves. The motion of an actuating piston can be changed quickly. A fluid system can give great flexibility in speed and motion control; it can give motion control in very small steps. Relief

Courtesy Mobile Aerial Towers, Inc.

Fig. 3. Aerial man lifts are powered hydraulically by closed-center, parallel circuit systems. An engine PTO drives a hydraulic pump which stores energy in an accumulator. The system operating pressure is 1,200 psi.

valves can easily be arranged to protect a system and to avoid damage. Control can be simple, efficient, and centralized. In general, fluid systems have relatively few moving mechanical parts; this means a high degree of reliability and a low maintenance cost.

Uses of Fluid Power

Where can fluid power be used? There is an old adage that states "wherever there is a smoke stack, there is likely to be an application for either air or hydraulic equipment." Today, the adage can be extended much further—nearly every industrial machine is somewhat dependent on the use of fluid power equipment. From the high-speed printing presses which print our daily newspapers, magazines, and telephone directories to the large semi-trailer trucks which bring our foodstuffs to the supermarkets, applications for air and hydraulic devices may be found. A person can walk down the main street of any town or city and discover that practically all the products in each store are processed by some type of air or hydraulic equipment.

Panorama of Fluid Power at Work*

A quick trip through a few industries basic to our country's economy can reveal the widespread use of fluid power. In mines and mills, farms and factories, warehouses and retail counters, construction projects and shipping terminals—at home and on the highway, and in defense and in public development—fluid power is at work.

Agriculture—Fluid power is used to lift the plow from its furrow and to lower it for the next furrow, as the farmer moves on his mechanized way. It operates the business ends of harrows, harvesters, hullers, planters, threshers, seeders, pickers, and diggers. Long before this equipment arrives at the farm, its manufacturer has made liberal use of fluid power in the shops and machines which have fashioned it.

Automotive—Power brakes, power steering, power windows, and powered seat adjustments are all typical fluid power devices. Automatic hydraulic transmissions rate as a high achievement of the art. At the production lines, fluid power is used to operate the presses which form body parts and fenders, to punch holes and heads of rivets which hold the frame together, to actuate heads, slides, and chucks of the machinery that is used to produce engine

---

*Prepared by the Education Committee of the National Fluid Power Association.

Fig. 4. On this hydraulic excavator, a hydraulic system automatically adjusts speed and power of the machine under all digging conditions. This is accomplished by an automatic regulating valve which varies the output of the pump—high speed in light digging conditions, high power in heavy digging conditions.

parts, to spin the wheels which polish the finishes, and to index "the line" itself.

Aviation—The pilot lifts his landing gear and controls ailerons, rudders, elevators, and trim tabs with fluid power. In the fighting aircraft, fluid power opens the bomb bays and rotates the turrets. "One-shot" casting operations for light aluminum and magnesium parts need fluid power to close the dies and to inject the metal. Wing and fuselage sections are formed in stretch presses that are operated by fluid power.

Construction—When it comes to moving dirt or rock—for roads, tunnels, dams, or seaways—look for fluid power at the member which bites the earth. It may be found on road graders, shovels, crushers, and rock drills. Rollers, scrapers, bulldozers, vibrator screens, draglines, truck loaders, and asphalt mixers all use fluid power. Even the smaller tools of construction get into the act—

17

heading of rivets, breaking of concrete, or jacking up of structural members.

Fig. 5. This loader has a hydraulic system set at 2,250 psi. There are two hydraulic pumps—main pump capacity 25 gpm and secondary pump capacity 10 gpm. The system has a 20-25 micron filter, and the maximum hydraulic lift capacity is 10,000 pounds. The loader is equipped with a rear-wheel 2,000 psi hydrostatic power steering 14:1 overall ratio.

Chemicals—Remote control of the thousands of valves, big and small, used throughout the chemical process industries comprises one of the more important newer applications of fluid power. It controls hopper gates and operates discharge doors on mixers and treating chambers. Fluid power regulates roll pressures in blending mills, packs chemicals into containers, and actuates panel controls.

Defense—Fluid power aims antiaircraft batteries, rotates tank turrets, catapults aircraft from flight decks, and elevates guided

Courtesy Caterpillar Tractor Company

Fig. 6. Large scrapers are employed in earth removal. Note the hydraulic equipment on this machine.

missiles to begin their grisly chores. On warships, fluid power functions from forecastle to rudder—steering, hoisting, positioning, and opening and closing operations. The service man of today—on land, on sea, or in the air—finds himself in a virtual fluid power laboratory.

Food Processing—Perhaps the most impressive use of fluid power in the food industry is in canning—from the moment the can is formed, until it is finally filled and labeled. Filling, capping, and sealing of all types of food containers receive a big assist from fluid power, as do the grinders, pulverizers, sifters, screeners, presses, cutters, dehydrators, conveyors, ovens, agitators, and wrapping and cartoning machinery.

Lumber—From the lusty cry of "TIMBER" to the finishing coat on fine furniture, fluid power lends its muscles to work in wood products. Logging grips clamp with either oil power or air power, as do the holding and feeding devices at the sawmill. Debarkers, conveyors, skidders, trimmers, and swing saws all utilize fluid power mechanisms. At the furniture factory, laminating, ve-

Fig. 7. This milling machine uses a large hydraulically operated clamping fixture to hold aluminum blower housings while the sides and ledges are being milled. The fixture has two hydraulic cylinders—one at either end, and is adjustable for holding seven different sizes of parts.

Courtesy Sundstrand Machine Tool

neering, sanding, gluing, and polishing, are often fluid power operations.

Materials Handling—The fork-lift trucks which move so efficiently and in such large fleets throughout almost all plants and warehouses these days are so versatile, because their telescoping masts and their grab jaws and pusher bars are fluid powered. Conveyors, hoists, cranes, dumpers, tilting ramps, and leveling docks are only a few machines in the roster of fluid powered tools for modern materials handling.

Marine—One of the classic jobs of fluid power in marine shipping is automatic helmsmanship. Cargo handling is partially a fluid power job. Even the hatch covers are actuated by fluid power. Barge and dredge operations on inland waterways are becoming more mechanized through fluid power. Canal and river lock routines make good use of fluid power for operating capstans, butterfly valves, winches, and hoists. Much of the dock and shipyard machinery is either hydraulic or pneumatic.

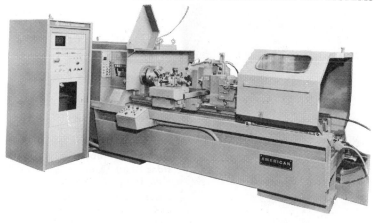

Courtesy American Tool Incorporated

Fig. 8. *Modern high-performance N/C turning center equipped with a power-operated chuck, either air or hydraulic, and an air-operated tailstock with 10 inches of spindle travel. The headstock spindle speed range is 30 to 3,000 rpm.*

Metalworking and Machine Tools—The basic machine tools of industry owe much of their productivity to the intelligent use of fluid power mechanisms. Controlling the table movements of a milling machine, feeding the cutting tools on an automatic lathe, advancing and feeding twist drills, driving drill presses and grinding wheels, actuating the ram of a broach, or clamping the work in the chucks and fixtures—these are all routine jobs for fluid power. The machine tool industry, indeed, is a large user of fluid power components.

Mining—A mainstay of modern mining operations is the voracious "mole," clawing out the earth's assets with fluid power actuated fingers. Throughout coal and mineral mining and quarrying, fluid powered devices are at work—from excavating operations to classifying, handling, and refining operations. Crushers, classifiers, shovels, cranes, hoists, and trucks make use of hydraulics. Portable air powered tools (chippers, drills, and hammers) play a large part, too, in the recovery industries.

Packaging—Fluid power is a common "muscle" in packaging machinery. It is used on scales, wrapping machinery, package

21

Courtesy W.F. & John Barnes, Babcock & Wilcox

Fig. 9. This 22-station Transfer Machine is specially designed for drilling, chamfering, reaming, and tapping V-8 or V-12 diesel cylinder heads. Hydraulic power is used throughout the machine to index the parts from station to station, to advance and return locators in the fixturing, to clamp the parts at each individual station, and to advance and return the feed units. Separate hydraulic power units can be seen along the periphery of the machine, complete with pump, motor, and tank. These units provide hydraulic oil at 500 psi to perform the above functions.

formers, carton fillers, tube and jar loaders, gluing machines, sealers, labelers, and bagging machines. Automatic tensioning of bands or straps is a natural application.

Courtesy Warner & Swasey Company

Fig. 10. This two-axis N/C turret lathe has a 12-station turret which handles ID and OD turning. The machine can be equipped with either an air-operated or a hydraulic-operated chuck up to 10 inches diameter, with a bar capacity of 2 inches. The machine can be used as either a chucker or a bar machine. The hydraulic system can have up to three pumps, depending on the usage.

Paper—In production of paper, the pulp must pass through many stages of continuous web flow. The amount of roll pressure that is exerted on the paper stock as it is consolidated from the mat to the finished stock is always critical. Controls for these pressures, together with speed control, are probably the most important applications of fluid power in papermaking. Calender roll adjustment is typical. Throughout the paper mills other fluid power devices may be found, such as drives, feeders, reels, dryers, shredders, vibrators, sizers, coating and laminating units, hydropulpers, and finishing tables.

Petroleum—From the raw crude oil to the finished fuel, oils, and greases, the piping of petroleum fluids, over long and short distances and under varying pressures and temperatures, depends on valving controls. Remote control of valves—sometimes there are acres of them in yards and tank farms, and sometimes there is only a single valve halfway across a desert—is a fluid power assignment, through cylinders or motors located at the valve stem. Fluid power systems also do their bit in extractors, draw works, drill transmissions, swivels, blocks, and pumping stations. Pneumatic instrumentation was pioneered at the refinery.

Plastics—Plastics fabrication is a molding business. Injection, preform, laminated, and vacuum molding pressures are all exerted and controlled by fluid power systems. On an injection machine, for example, the dies are closed and the plastic material is forced into the die cavities by fluid power. Extrusion presses for tubular plastic products, such as garden hose, also use fluid power for the "push" action.

Printing—Modern printing presses operate at lightning speeds. These speeds are adjustable to close tolerances over wide ranges, and respond instantly to push-button cues. Fluid power variable-

Courtesy The Cross Company

Fig. 11. Overall view of a hydraulically operated machine that automatically assembles and tests transmission stator support assemblies. Note the large hydraulic power unit at the left.

Courtesy National Automatic Tool Co.

Fig. 12. Large 18-spindle drilling machines with a production rate of up to 1,028 gross per hour. Each machine is equipped with a 36-in. diameter, four-position rotary table. Parts are manually loaded and automatically clamped. Note the hydraulic clamping cylinders on the fixtures.

Courtesy Gilman Engineering & Manufacturing Co.

Fig. 13. Overall view of a piston, pin, and rod automatic assembly machine. This 12-station, 9-in. Transferline machine assembles piston, wrist pin, and connecting rods at the rate of 1,200 per hour gross. Note the pneumatic equipment employed.

25

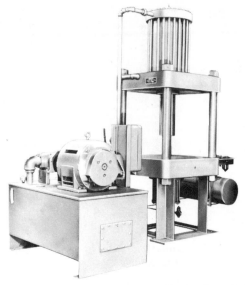

Courtesy Logansport Machine Co., Inc.

Fig. 14. Large forming press using hydraulic power as the motivating means. Note the large cylinder mounted on top of the upper platen. Also, note the heavy-duty hydraulic piping. Press is equipped with a die cushion cylinder, air-operated, under the lower platen. Air storage tank is located on the side of the press.

speed drives and controls make this possible. Inching, accelerating, decelerating, braking, roll pressure, and unwind and rewind tensioning are accurately performed with fluid power.

Fluid power has eliminated jerking, which is one cause of occasional off-register pictures that were seen in earlier color printing jobs.

Railroading—Railroads use fluid power in many ways: in the brakes that insure safety; in the shock system which smooths the ride; and in the new doors which open at the touch of a finger. The roadbed that anchors the ties and track is regularly tamped by machines with crab-like clusters of feet worked and controlled by fluid power. The very machines which handle, clean, and place the roadbed ballast make equally good use of fluid power. As for the locomotive—just peek inside a modern diesel.

Rubber—Fluid power is seen in many places in the rubber industry—in the production of new synthetics, in compounding, and

Courtesy Industrial Metal Products Corporation

Fig. 15. Automatic roll straightening machine on which parts are straightened from 0.060 TIR to 0.003 TIR. Production is 1,600 parts per hour gross. Parts are loaded on a chain-type transfer from a hopper unit and traversed into the straightening station by means of hydraulically operated fingers. The precision hydraulic cam-controlled straightening ram straightens four parts at a time. The parts are automatically unloaded by a separate chain transfer.

Courtesy Giddings & Lewis Machine Tool Company

Fig. 16. Precision machining center uses much hydraulic equipment. The shuttle pallet system permits setup during the machine cycle. This reduces setup time to the time needed to exchange pallets. Machine is equipped with an automatic tool-changer.

27

Fig. 17. This 16,000,000-lb hydraulic stretcher processes aluminum plate up to 6 inches thickness at Alcoa's Davenport, Iowa, sheet and plate mill. The stretching process relieves internal stresses and subsequently permits intricate machining with no appreciable warpage. This machine is operated by Alcoa for the United States Air Force and is one of the largest of its kind in the world.

in the fabrication of finished products. Roll pressure control during compounding is a major need (squeezing each batch time and again between the rolls of a mixer). All the molding machines are fluid power operated. Cutting, dipping, forming, molding, extruding, vulcanizing, threading, calendering, turning, and finishing— all are rubber industry processes where pressures or motions are regulated by fluid power.

Steel—The enormous pressures required to knead a steel ingot through the dozens of successive shapes, until a fine gleaming sheet or a flawless reel of wire is obtained, require the huskiest power cylinders—and the most precise. Throughout these stages, both hot and cold, the control of roll pressure is maintained by fluid power. One can mention many more uses for fluid power by the steelmaker—rod and tube mill control, furnace push-out and door-raising operations, draw bench pulls, scale breaker operation, and the manipulation of the giant ladles of white-hot metal which

Courtesy The Cross Company

Fig. 18. View of hydraulically operated cylinder block definning machine used in high-production automated foundries. Note the hydraulic power units and controls mounted along the rear of this machine.

are foremost in every portrayal of steelmaking's most dramatic scene.

Textile—The striking colors and patterns worn by young ladies (and in sports shirts worn by men) owe their popularity to the modern, automatic textile printer—which uses fluid power for speed regulation, plate pressure, and color feed. Down the long road that leads from raw fiber, natural or synthetic, to the finished article, fluid power plays its part—in squeezers, balers, beam warpers, washers, roving and drawing frames, slitters, winders and rewinders, and folders.

REVIEW QUESTIONS

1. Hydraulic brakes are very common on passenger automobiles. Explain their use.

29

2. Explain why pneumatic devices, rather than electrical devices, are used on an operation, such as digging a tunnel under a river.

3. A relief valve is found on the home hot-water heater. Explain the use of this valve.

4. Pneumatic devices are used for road and street repair. List several reasons why compressed air is used for this type of operation.

5. Explain the increased use of fluid power devices in the past decade.

6. List some examples of the use of fluid power equipment in processing food.

CHAPTER 2

Pressure, Work, and Power

Basic terms that are commonly used in the field of hydraulics and pneumatics must be discussed and understood.

PRESSURE

The word *pressure* is defined as *force per unit area*. Although other units may be used, pressure is commonly expressed in such units as pounds per square inch. The abbreviation *psi* is usually employed to indicate *pounds per square inch*.

Fig. 1 shows an arrangement of two cylinders that are connected by a pipe or tube. A close-fitting piston is placed in each cylinder. In each cylinder (under the piston), the liquid and the connecting tube are shown. If it is assumed that there is no movement of each piston and that there is no leakage past each piston, the liquid and all the parts are at rest—a static condition. It is also assumed that a force F_1 of 100 pounds acts on piston No. 1 and that there is no friction between each piston and its cylinder wall. If piston No. 1 has a flat or face area of 2 square inches that is in direct contact with the liquid, the pressure in the liquid under piston No. 1 is equal to the *force* divided by the *area* (100 divided by 2), or 50 pounds per square inch. Thus, the liquid pressure at the face of piston No. 1 is 50 *psi*.

Assuming that piston No. 2 is essentially at the same level as piston No. 1, the liquid between the pistons serves as a medium to transmit the pressure from one piston face to the other piston face. Thus, the liquid pressure at the face of piston No. 2 is 50 *psi*. If the area of piston No. 2 is 6 square inches, the force F_2 on the face of piston No. 1 is (6×50), or 300 pounds. Thus, a force

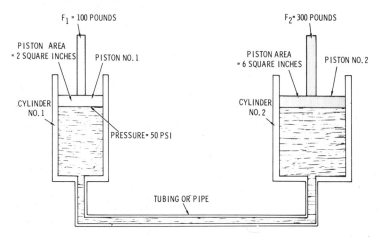

Fig. 1. Illustrating a hydraulic system having two cylinders and two pistons of different sizes.

of 100 pounds at piston No. 1 develops a force of 300 pounds at piston No. 2; this is accomplished by making the area of piston No. 2 equal to three times the area of piston No. 1. In a sense, the arrangement (see Fig. 1) is a fluid lever, similar to a mechanical lever using a metal bar and pivot.

Equal pressure at every point and in every direction in the body of a static liquid (a liquid at rest) is characteristic of all static fluids, liquids, or gases. This is called *Pascal's law,* after an early experimenter in this field of study. This law of pressure is very useful, and can be used to advantage in countless applications.

Atmospheric Pressure

A blanket of air surrounds the earth; this is called the *atmosphere.* At the surface of the earth, *atmospheric pressure,* which is due to the weight of the air above the surface of the earth, can be measured. Atmospheric pressure is commonly measured with a mercury barometer. Thus, atmospheric pressure is often called *barometric pressure.*

Fig. 2 illustrates the basic principle of a mercury barometer. The glass tube is open at the lower end and closed at the upper end. Initially, the tube is completely filled with pure mercury; then it is inverted, with the open end submerged, in a small ves-

32

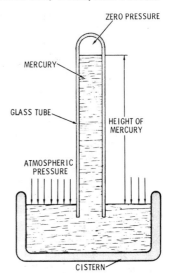

Fig. 2. Illustrating the basic parts of a mercury barometer.

sel or cistern containing mercury. The height of the column of mercury gives the direct reading of the barometer; the weight of the air above the barometer balances the weight of the mercury column. Barometric pressure is usually expressed in inches of mercury. A barometric height of 29.92 inches of mercury corresponds to an atmospheric pressure of about 14.7 pounds per square inch.

Pressure Measurement

Many instruments or gauges that are used for measuring pressure employ a Bourdon tube. Fig. 3 illustrates this type of gauge. The Bourdon tube is a hollow metal tube that is made of brass or a similar material; it is oval or elliptical in cross section, and is bent in the form of a circle. One end of the Bourdon tube is fixed to the frame at point A (where the fluid enters) the other end B (closed) is free to move. The free end B actuates a pointer through a linkage system. As fluid pressure inside the tube changes, the elliptical cross section changes, and the free end B of the Bourdon tube moves inward or outward, depending on the character of the change. A convenient pressure scale or dial can be arranged from a calibration of the gauge.

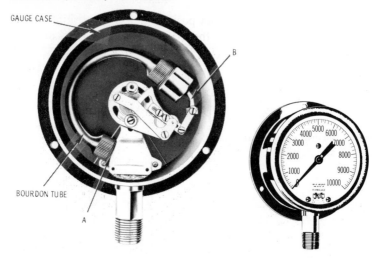

Fig. 3. Cutaway view of a Bourdon tube type of pressure gauge (left), and the indicator dial (right).

The position of the free end *B* of the Bourdon tube depends on the difference in fluid pressure between the inside and the outside of the tube. If the outside of the Bourdon tube is exposed to atmospheric air pressure, the instrument reading is a measure of so-called *gauge pressure*. For example, if the pressure reading at the outlet of an air compressor is 100 *psi gauge*, this indicates that the outlet air pressure is 100 *psi* above atmospheric pressure. In this book, all pressures referred to are *gauge* pressures.

In some instances, the pressure in a piece of equipment may be below atmospheric pressure; this condition is designated as *vacuum.* For example, if the air in a tank is at a pressure that is below atmospheric pressure, the pressure gauge indicates a certain *vacuum,* or negative gauge pressure.

DEFINITION OF WORK, ENERGY, AND POWER

As shown in Fig. 4, a body weighing 20 pounds at a given level is indicated in position No. 1. If the body is moved vertically through a distance or displacement of 9 feet, the action involves *work*. The technical term *work* is defined as the product of *force times displacement,* with the force in the direction of the displacement. As the body moves from position No. 1 to position No. 2, a force of 20 pounds moves the 20-lb. body through a displacement of 9 feet. This equals (9×20), or 180 foot-pounds of work.

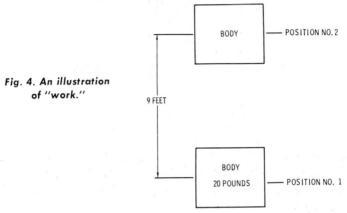

Fig. 4. An illustration of "work."

Energy is defined as the *capacity to do work.* Energy refers to a possibility. A body resting at position No. 2 has a certain capacity, or a certain energy. If the body is moved to the level 9 feet below, (9×20), or 180 foot-pounds is available to do work.

The term *work* in itself does not involve a time element. Rate of movement, or *speed,* is often important. *Power* is defined as the *time rate of doing work.* If the body weighing 20 pounds were moved at a constant speed and in a vertical direction through a vertical distance of 9 feet in a time of 2 seconds, the "power" can be calculated as follows:

$$\text{power} = \frac{20 \text{ pounds} \times 9 \text{ feet}}{2 \text{ seconds}} = \frac{90 \text{ foot-pounds}}{\text{second}}$$

Thus, the power required is 90 foot-pounds per second. *One horsepower* has been arbitrarily defined as equal to *550 foot-pounds per second.*

FORCE AND WORK IN A FLUID DEVICE

Fig. 5 is an illustration of a pump or compressor delivering fluid (either oil or compressed air) to the left-hand side of a piston in a cylinder. Let P represent the fluid pressure, in *psi*, and A represent the piston area in *sq. in.* (abbreviation for square inches).

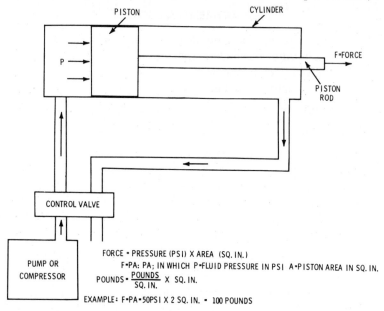

Fig. 5. *Illustrating the relation between pressure, area, and force.*

Then the force F acting on the left-hand face of the piston is PA. For a pressure P of 50 *psi* and an area A of 2 *sq. in.*, the force acting on the left-hand face of the piston is equal to (50×2), or 100 pounds. Assuming no friction due to the cylinder wall, the force F at the piston rod is equal to 100 pounds.

Fig. 6 is another illustration of a pump or compressor delivering fluid to the left-hand side of a piston in a cylinder. As in the previous example, let P represent the fluid pressure (*psi*) and let A represent the left-hand piston area (*sq. in.*). Then the fluid force F acting on the left-hand piston face is $F = PA$. If this force remains constant while pushing the piston through a displacement or distance L (inches), the work done by the fluid on the left-hand face of the piston is equal to the force F times the displacement L or the work $W = PAL = FL$. For a fluid pressure of 50 *psi*, a piston area of 2 *sq. in.*, and a displacement of 3 *inches*, the work W can be determined as follows:

$$W = (50 \times 2 \times 3) = 300 \ inch\text{-}pounds$$

DISPLACEMENT ACTION

Fig. 7 is a diagram of a pump delivering hydraulic fluid through a pipe or tube into a cylinder. The piston is shown at position No. 1 at a given time or part of the stroke. If the pump continues to deliver fluid into the cylinder, the fluid pushes the piston to the right-hand side a distance of 3 *inches* (position No. 2). The linear displacement or movement of the piston to the right is equal to 3 *inches*. For a piston area of 2 *square inches* and a piston displacement of 3 *inches,* the *volumetric displacement* of the piston is equal to (2 *sq. in.* × 3 *in.*), or 6 *cubic inches.* Assuming no leakage of fluid across the piston from the left-hand side of the piston to the right-hand side of the piston, the total amount of fluid added to the cylinder is equal to 6 *cubic inches;* in other words, 6 *cubic inches* of fluid was admitted to the cylinder and

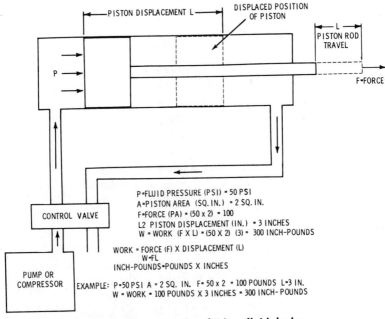

Fig. 6. Illustrating "work" in a fluid device.

pushed the piston, for a *volumetric displacement* equal to 6 cubic *inches*. With no leakage, this action is frequently called a *positive-displacement* action. The amount of fluid entering the cylinder is equal to the *volumetric displacement* of the piston.

In reference to the system illustrated in Fig. 1 (two cylinders and two pistons), the area of piston No. 1 is 2 *square inches* and the area of piston No. 2 is 6 *square inches* (3 times that of piston No. 1). If a force of 100 *pounds* is applied at piston No. 1,

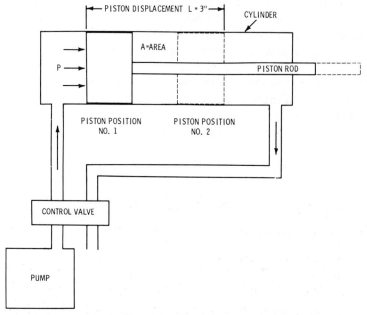

Fig. 7. Illustrating displacement action.

there is a corresponding force of 300 *pounds* at piston No. 2. For a given fluid pressure, there is a force multiplication because of the difference in piston areas, assuming no leakage. If it is assumed that piston No. 1 moves downward a distance of 0.03 *inch* for a *positive-displacement* action of the incompressible hydraulic fluid, the volume of fluid displaced by piston No. 1 is equal to the volume of fluid displaced by piston No. 2. For a *positive-displacement* action, piston No. 2 then moves upward 0.01 *inch* (piston area times displacement is equal to the displaced volume). For piston No. 1, the total *work* is (100×0.03), or 3 *inch-pounds*. For piston No. 2, the total *work* is (300×0.01), or 3 *inch-pounds*.

Rate of Flow and Piston Travel

The volume rate of fluid flow through a device can be expressed in various units. For example, it is common to express *volume rate of flow* for liquids in gallons per minute (*gpm*). One gallon is equal to 231 *cubic inches*. For a practical example, an oil pump may be said to deliver a flow of 10 *gpm*; this corresponds to a rate of (10×231), or 2310 *cubic inches per minute*. For air flow, it is common to use a volume rate stated as cubic feet per minute (*cfm*). For example, it may be said that air enters an air compressor at a rate of 30 *cubic feet per minute*. The term *capacity* is used to designate the *volume rate of flow* through a pump.

Fig. 8A illustrates the flow of hydraulic liquid from a pump. A volume rate Q of 3 *gpm* indicates a flow rate of 11.55 *cubic inches per second*. As illustrated in Fig. 8B, the piston position has moved from position No. 1 to position No. 2, and there is a given constant volume rate of flow Q from the pump. Let A represent the face area of the piston (*square inches*) and V_p the constant velocity or speed of the piston (*inches per second*). Assume no leakage of fluid past the piston; in other words, there is a *positive-displacement action*. During a given time there is a given volume rate of flow from the pump (*cubic inches*). The volume leaving the pump must equal the *volumetric displacement* of the piston. The constant volume rate Q is the time rate of pump delivery. The product AV_p is equal to the time rate of cylinder volume displacement, if there is no leakage of fluid, as shown by the formula: $Q = AV_p$.

If the volume rate Q is equal to 3 *gpm*, the time rate of cylinder volume displacement corresponds to 11.55 cubic inches per second. If the piston face area is 5 *square inches*, the speed of the piston is equal to:

$$V_p = Q/A = 11.55/5 = 2.31 \text{ inches per second}$$

Thus, the speed of the piston depends on the volume rate of delivery from the pump.

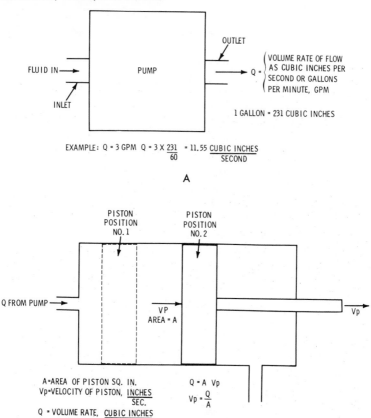

Fig. 8. Illustrating flow of hydraulic liquid from a pump: (A) Pump outlet flow; (B) Rate of piston travel.

SUMMARY

When a fluid acts on one side of a piston, as illustrated in Fig. 7, let P represent the differential fluid pressure across the piston (in *psi*) and A the area (*sq. in.*). The fluid force F (in *pounds*) on the piston can be determined by the formula: $F = PA$. Other units can be used; the foregoing is used to illustrate a common set of units.

Let L represent linear displacement or movement of the piston (in *feet*) to the right with a constant force F (see Fig. 7). For the movement from position No. 1 to position No. 2, the work W, in *foot-pounds*, is given by the relation: $W = FL$.

As the piston moves from position No. 1 to position No. 2 during the time interval t (in *seconds*), assume that the piston velocity V_p (in *feet per second*) is constant. The piston velocity V_p can be expressed in the form: $V_p = L/t$. During the displacement from position No. 1 to position No. 2, with no leakage, the volume rate of flow Q is given by the relation: $Q = AV_p$. If A is given in *square feet* and V_p is given in *feet per second*, then Q is in terms of *cubic feet per second*. Various conversions are possible. As indicated in the Appendix—Conversion Factors, for example, 1 gallon per minute (*gpm*) is equal to 0.002228 *cubic feet per second.*

Power is defined as the time rate of doing work. Assuming a constant linear velocity V_p as the piston moves from position No. 1 to position No. 2, let R represent power (*foot-pounds per second*). Then, power R can be expressed in the following forms as:

$$R = FL/t = FV_p; \text{ and } R = PAV_p = PQ$$

The first power equation shows that *power* is proportional to the product of *force* times *piston velocity*. As shown in the second equation, *power* can also be expressed as the product of the *pressure difference* and the *volume rate of flow*.

As an example, if the volume rate of flow Q is 3 *gpm* (corresponding to 11.55 *cubic inches per second*) and the pressure differential is 500 *psi,* then the power becomes:

$$R = 500 \times 11.55 = 5775 \text{ inch-pounds per second}$$

$$R = (500 \times 11.55)/12 = 481 \text{ foot-pounds per second}$$

Since 1 *horsepower* equals 550 *foot-pounds per second*, the power becomes:

$$R = (500 \times 11.55)/(12 \times 550) = 0.875 \; horsepower$$

The term *torque* is used in connection with turning and twist-ing actions. An automotive internal combustion engine develops a torque at the crankshaft to turn the crankshaft and ultimately to turn the rear wheels of the car. An electric motor develops a torque at a rotating shaft. As illustrated in Fig. 9, if a motion in

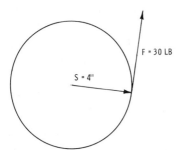

Fig. 9. Illustration of "torque."

a circular path has a radius S of 4 *inches* at which there is a tangential force F of 30 *pounds* acting at right angles to the radial line, the "torque" can be defined as the product FS; in this ex-ample, the torque is equal to (30×4), or 120 *inch-pounds*.

Assuming a circular path for a radius S for one complete revo-lution, the distance involved is $2\pi S$, where π is equal to approxi-mately 3.14. For an angle speed of N revolutions per unit time, the linear velocity V (speed of a point moving in a circular path) is given by the relation:

$$V = 2\pi SN$$

For example, for a radius of 4 *inches* and a speed of 200 *rpm* (revolutions per minute), the linear velocity V is equal to:

$$V = 2\pi \, (4 \times 200) = 5020 \; inches \; per \; minute$$

$$V = (2\pi 4 \times 200)/(12 \times 60) = 6.98 \; feet \; per \; second$$

REVIEW QUESTIONS

1. What is the purpose of the oil gauge on the dash panel of an automobile?

2. If a man walks up a flight of stairs (a vertical distance of 20 feet), how much work does he do, assuming his weight to be 170 pounds?

3. In question No. 2, if a time interval of 20 seconds is required for the man to walk up the stairs, what is the horsepower expended?

4. If the piston area in a forge press is 4 square inches, and it is desired to develop a piston force of 1 ton, what oil pressure is needed, neglecting friction?

5. In question No. 4, neglecting leakage, it is desired to move the piston at a speed of 5 feet per second. What capacity, in gallons per minute, must the pump deliver to the cylinder?

6. A torque wrench is a wrench that indicates the torque developed. Explain how a torque wrench can be of value in tightening a screw or bolt.

7. What is the difference between pressure and force?

8. What is the difference between work and torque?

CHAPTER 3

General Features of Machines

A wide variety of devices are used in industrial practices. Familiarity with certain basic or common features of these devices can aid in gaining a better understanding of hydraulic and pneumatic equipment, and can be of help in installation, operation, and maintenance of the equipment. This chapter discusses some of these basic features and also defines certain terms that are used in subsequent chapters to explain the fundamental operation of various fluid devices.

MACHINES AND MECHANISMS

As discussed in the previous chapter, the term *work* is defined as the product of force times distance (or displacement), with the force in the direction of the distance. If a body weighing 10 pounds is lifted vertically through a distance of 2 feet, work has been performed. The amount of work that has been done is equal to (2×10), or 20 foot-pounds. *Energy* is the ability or capacity to do work.

In a machine, energy can be converted from one form of energy to another. A machine is a device for doing work. For example, an electric motor converts electrical energy to mechanical energy. Electrical energy enters the motor by means of wires; the motor output (mechanical energy) is delivered by means of a shaft which is rotated when the current is turned on. This shaft can be connected to various devices to exert a force which moves through a certain distance; in other words, the output shaft does work. For another example, a generating station for electricity consists of electric generators. In the electrical generator, a rotating shaft is turned by a mechanical means (either steam or hydraulic turbine). Therefore, mechanical energy is converted to electrical energy.

Industrial devices can be classified generally into two groups: (1) machines; and (2) mechanisms. The word *mechanism* indicates a device primarily involving only motion. For example, the motor in a household refrigerator differs from a spring in a watch in that the spring in the watch overcomes a small amount of friction, although the main purpose is to achieve a definite motion. The watch does no useful work in the sense that a force moves a load through a distance; the watch is classified as a "mechanism," whereas the refrigerator motor is classified as a "machine." The term *mechanism* is used in connection with the various *control devices*. In fluid power work, both machines and mechanisms are of importance.

FLUID MACHINES

In the electric generator, mechanical energy is converted to electrical energy. In the *fluid pump*, mechanical energy is converted to fluid energy. The electric generator and pump are thus similar.

Fluid Pumps

Fig. 1 illustrates the action of a fluid pump. The piston exerts a force on the fluid. If the piston moves through a distance, the piston does mechanical work on the fluid. This action increases

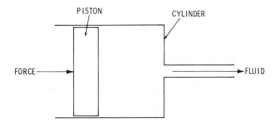

Fig. 1. Illustrating action of a fluid pump.

the pressure of the fluid. Thus, mechanical energy is converted to fluid pressure energy.

Fluid Motors

In the electric motor, electrical energy is converted to mechanical work. In the *fluid motor,* fluid energy is converted to mechani-

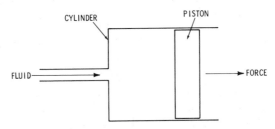

Fig. 2. Illustrating action of a fluid motor.

cal work. Therefore, the electric motor and the fluid motor are similar. A fluid motor in action is illustrated in Fig. 2. Fluid pressure energy acts on the piston. The piston, in turn, exerts a force on a load. If the piston moves through a distance, the piston then does work.

Figs. 1 and 2 illustrate the motion of a piston in a straight line. If the piston moves back and forth, the motion is *reciprocating.* In most automobile engines, the pistons move back and forth, which is a reciprocating action. The engine in a car is an example of a *motor;* in this type of motor, energy from the fuel (gasoline) is converted to mechanical work. Also in the typical automobile engine, the crankshaft rotates, which is a typical example of rotary motion.

A fluid pump may involve either a reciprocating action or a rotary action; likewise, a fluid motor can involve either a reciprocating or a rotary action.

GENERAL TYPES OF FLUID MACHINES

Fluid machines are divided into two main types, depending on the type of fluid action: (1) the velocity, or dynamic, type and (2) the positive-displacement, or pressure, type. In the velocity, or dynamic, type of fluid machine, the action between a mechanical part and the fluid involves appreciable changes in velocity. The

47

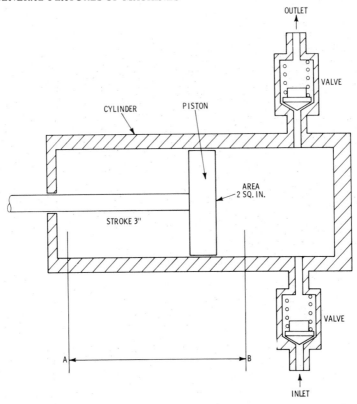

Fig. 3. Schematic diagram of a machine having reciprocating action.

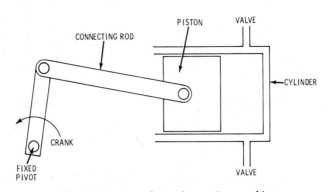

Fig. 4. Diagram of a reciprocating machine.

common propeller type of desk or ventilating fan is an example of the velocity type. In the positive-displacement, or pressure, type of fluid machine, the characteristic action is a volumetric

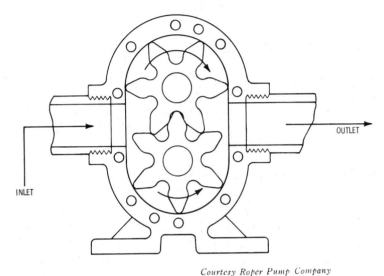

Courtesy Roper Pump Company

Fig. 5. Diagram of a rotary type of machine, showing rotary gear.

change or displacement action. Pressure is developed primarily by a displacement action. In this text, attention is placed primarily on positive-displacement machines; these are more commonly used in hydraulic and pneumatic equipment. An example of a displacement action is found in Fig. 7 of Chapter 2.

The general construction of positive-displacement machines is divided into the following two groups, depending on the type of mechanical motion: (1) *reciprocating* machines; and (2) *rotary* machines.

Reciprocating Machines

A schematic diagram of a reciprocating piston machine is shown in Fig. 3. The machine consists of a cylinder which includes a reciprocating plunger or piston, and valves for directing the fluid to and from the cylinder; this is also an example of the action of

49

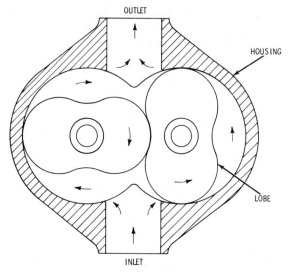

Fig. 6. Diagram of a rotary lobe-type machine, showing rotary lobe.

a pump. Fluid enters the inlet valve as the piston moves to the left-hand side; during this action the outlet valve is closed. For the stroke to the right-hand side, the inlet valve is closed, the outlet valve is opened, and the piston pushes fluid through the outlet valve. When the machine uses a liquid, such as oil, it is usually called a *pump*—when the machine uses either gas or air, it is usually called a *compressor*.

If, for example, the area of the piston is equal to 2 square inches and the stroke of the piston from point A, to point B is 3 inches (stroke is the maximum travel of the piston), the "volumetric displacement" or simply the "displacement" is equal to (2×3), or 6 cubic inches, assuming that there is no leakage of oil past the piston. Thus, as the piston moves from position A to position B, 6 cubic inches of oil are pushed through the outlet valve; it can be said that the oil delivery of the machine is 6 cubic inches. The volume rate of flow is equal to the volume per unit of time. If 6 cubic inches of oil are delivered in a time of 1 minute, the volume rate of flow is equal to 6 cubic inches per minute; if the 6 cubic inches were delivered in a time of 2 minutes, the volume rate of flow would then be equal to 6/2, or 3 cubic inches per minute. In actual practice, for hydraulic machines, volume rate of flow is commonly expressed in gallons per unit of time—one gallon occupies a volume of 231 cubic inches.

A positive-displacement machine may be either a fixed-displacement type or a variable-displacement type, depending on the arrangement of its parts. In the fixed-displacement machine, the arrangement of parts *cannot be changed* to vary the displacement; in the variable-displacement machine, the arrangements of parts *can be changed* to vary the displacement. For example, in Fig. 4, the crank rotates about a fixed pivot. A connecting rod connects one end of the crank with the piston. In this type of reciprocating machine, the stroke of the piston cannot be changed. For each revolution of the crank there is a complete stroke of the piston and, in turn, fixed displacement. The automobile engine has a fixed displacement or given number of cubic inches for each revolution of the crank. If the stroke could be changed, the displacement would then be variable. The stroke can be varied, for example, by arranging for a change in the length of the crank, as in some types of fluid pumps.

Rotary Machines

In a rotary machine, there is a rotary motion about an axis. Rotary machines are usually classified with respect to the machine part or mechanical element that rotates.

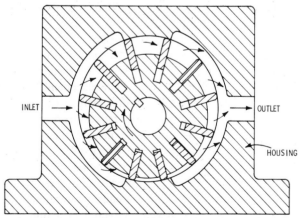

Fig. 7. Diagram of a rotary type of machine, showing the rotary vane.

51

A rotary gear machine is illustrated in Fig. 5. A pair of meshed gears is enclosed in a housing. As the gears rotate, the fluid is trapped between the gear teeth and the housing, and is carried around from the inlet to the outlet. During each revolution of the gears, a volume of fluid is transferred from the inlet to the outlet. This is a fixed-displacement machine. Mechanical energy can be applied to the gears to force fluid through the machine; then the machine can be used as a pump or compressor. Fluid under pressure can be forced into the machine to drive the gears and to perform mechanical work at the gear shaft; then the machine becomes a fluid gear motor.

A rotary lobe-type machine is shown in Fig. 6. This machine differs from the rotary gear type in that lobed elements are used, rather than gears; it can be used for both oil and air, and as a pump, compressor, or fluid motor.

Fig. 7 is an illustration of a rotary vane type of machine. The rotating member, with its sliding vanes, is set off center in the housing. The entering fluid is trapped between the vanes (which ride on the inside of the housing) and is carried to the outlet. The rotary vane machine is used as a pump, compressor, and fluid motor. The eccentricity can be defined as the distance between the center of the rotating member and the center of the fixed housing. If the eccentricity is fixed, the fluid displacement or volume rate of flow is also fixed. If the eccentricity is variable, the fluid displacement or volume rate of flow can be varied or changed. A rotary machine may have either the radial-type or the axial-type pistons.

Fig. 8 illustrates the arrangement of the pistons in a radial piston pump; there is also a fixed guide or reaction ring. The pistons move back and forth in the cylinder, and are arranged in a radial direction. The cylinder rotates about a fixed spindle or pintle in which the valve ports are located. For pump action, the cylinder, which is eccentric with respect to the guide axis, is rotated by some mechanical means. The pistons move in and out radially; they travel outward on the inlet or suction stroke and inward on the discharge stroke. The eccentricity between the cylinder and the guide ring can be varied to give variable delivery. This arrangement can be used as a fluid motor.

A rotary pump with an axial piston arrangement is diagrammed in Fig. 9. The piston reciprocates in a direction parallel with the axis of rotation of the cylinder. In the pump, a drive shaft rotates the pistons and the cylinder. A stationary valve plate faces the cylinder. The drive shaft axis is located at an angle with respect to the axis of the cylinder block. As the drive shaft rotates, the cylinder block and pistons also rotate. Each piston moves in the direction of the valve plate during one-half of the drive shaft revolution and away from the valve plate during the other one-half of the revolution. In the pump, the fluid enters the cylinder as the piston is moving away from the valve plate.

The displacement of the pump with axial piston arrangement can be varied by changing the angle between the axis of the drive shaft and the axis of the cylinder. Thus the angle can be varied to give variable delivery.

Actuators

In a general sense, the term *actuator* can be defined as a device that converts fluid energy to mechanical motion. A wide variety of actuators is available. Actuators may involve both linear motion and/or rotary motion.

Actuating cylinders include one or more pistons and piston rods with the necessary seals. In a double-acting cylinder, fluid under pressure can be applied to either side of the piston to provide movement in the corresponding direction. In a single-acting cylinder, fluid under pressure can be applied to only one side of the piston, and a spring or gravity is used to return the piston to its original position.

A number of features are illustrated in Fig. 10. In the tandem actuators, two cylinders are mounted on a structural frame, and the piston rods are coupled together so that the piston assemblies function as a single element. Fluid connections are located at both ends of each cylinder. Fluid under pressure acts on the pistons to provide linear motion of a particular point on the member connecting the piston rods. As noted in the illustration, the lever oscillates about a pivot point. A roller (on the piston rod connector) rides in a slot in the lever. Linear motion of the piston rods results in a rotary motion of the lever. A quad arrangement of actuators, in which there are four cylinders, is illustrated in

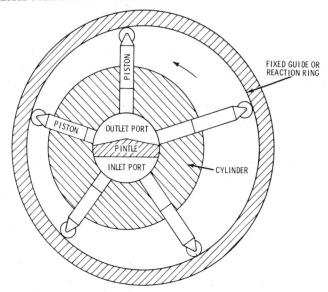

Fig. 8. Schematic diagram of a rotary type of pump with a radial piston arrangement.

Fig. 11. Two tandem actuators are mounted on a bracket; a double-slotted bar lever is extended in such a manner that the four cylinders provide linear motion, giving rotary motion to the lever. In the quad actuator mounted on a plug valve (Fig. 12), four cylinders with openings for fluid connections on the outboard end of each cylinder are shown. The actuator, because of the linear motion of each piston rod, provides a rotary motion to the valve, opening or closing the valve as desired. Actuators of this type may be operated by air, gas, or hydraulic fluids ranging from 50 to 2000 *psi*.

In various types of rotary actuators, a shaft is rotated through either a fixed or an adjustable arc, using a source of fluid under pressure. In one general type of actuator, for example, one or more vanes are attached to the drive shaft. The vanes are sealed in a cylindrical chamber, and oscillate between stops located in the housing. The actuator is rotated by a differential pressure across the vanes. In another general type of actuator, a helix is machined in the drive shaft, meshing with a helix in a piston. As

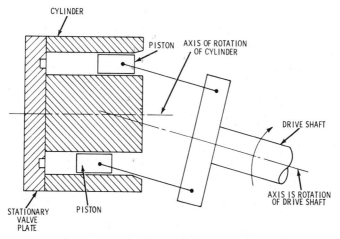

Fig. 9. Schematic diagram of a rotary type of pump with an axial
piston arrangement.

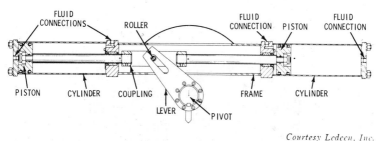

Courtesy Ledeen, Inc.

Fig. 10. Diagram of tandem actuators.

the piston moves back and forth, the shaft oscillates. Guide rods
pass through holes in the piston to keep it from rotating. In still
another general type of actuator, a rack, or racks, rotates a pinion
attached to the drive shaft. The rack connects to the pistons which
move back and forth to rotate the drive shaft.

PERFORMANCE FEATURES OF SOME MACHINES

Performance characteristics representative of commercially available products are given in Tables 1, 2, 3, and 4. More detailed
information can be found in the catalogs of specific manufacturers.

55

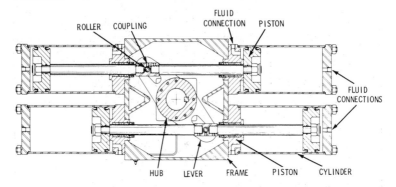

Courtesy Ledeen, Inc.

Fig. 11. Diagram of quad actuators.

These tables are presented to give an idea of the magnitude of machines that are used in current practice. In the tables, the following abbreviations are used:

Courtesy Ledeen, Inc.

Fig. 12. A quad actuator mounted on a plug valve. Note the four cylinders in the device.

psi = pounds per square inch
rpm = revolutions per minute
cfm = cubic feet of air at atmospheric conditions entering the compressor per minute
hp = horsepower

Table 1. Air Compressors

Types	Rotary and Reciprocating
Maximum pressure	50 to over 1000 *psi*
Speed	100 to over 2000 *rpm*
Capacity	5 to over 1000 *cfm*

Table 2. Cylinders

	CYLINDERS	
	Air	Hydraulic
Bore diameter	under 1 in. to over 20 in.	under 1 in. to over 16 in.
Maximum operating pressure	90 to over 250 *psi*	250 to over 5000 *psi*

Table 3. Motors

Motors, Air
Types	Axial-piston, radial-piston, vane, gear, lobe
Maximum operating pressure	50 to over 1000 *psi*
Maximum shaft speed	500 to over 5000 *rpm*
Output	0 to 10 *hp* or more

Motors, Hydraulic, Vane-Type
Maximum operating pressure	1000 to over 2000 *psi*
Shaft speeds	under 50 to over 2000 *rpm*
Maximum output	5 to over 60 *hp*

Motors, Hydraulic, Gear-Type
Maximum operating pressure	1000 to over 5000 *psi*
Maximum shaft speed	600 to over 2000 *rpm*
Maximum output	5 to over 60 *hp*

Motors, Hydraulic, Piston-Type
Maximum operating pressure	1000 to over 3000 *psi*
Shaft speed	under 50 to over 2000 *rpm*
Maximum output	5 to over 80 *hp*

Table 4. Pumps

Type of Pump	Maximum pressure (psi)	Maximum speed (rpm)
Gear	500 to over 2000	600 to over 3600
Piston	1000 to over 10,000	600 to over 180
Vane	1000 to over 2000	1200 to over 3600

CAVITATION

As a liquid flows through a passage or a machine, an undesirable action called *cavitation* may take place if certain conditions prevail. The word "cavitation" itself implies a cavity or void.

If, at some point in the liquid, the fluid pressure is low enough (below vaporizing pressure), the liquid may boil or form a vapor bubble or cavity. If the fluid pressure fluctuates both below and above the vaporizing pressure, alternating formation and collapse of the vapor bubbles occur. The violent collapse (occurring in a very short period of time) of vapor bubbles can force the liquid at a high velocity into the vapor-filled pores of the metal. The sudden halting at the bottom of the pore can produce high surge pressures on small areas. If these surge pressures exceed the strength of the metal, metal particles may be blasted out, the metal developing a spongy appearance. Because of cavitation, a fine suspension of metallic dust in the originally clean hydraulic fluid may present a serious hazard to proper system operation.

In addition to the parts becoming pitted, cavitation can be the cause of a drop in performance of a liquid pump, motor, or other device. For example, dissolved air or gas in the hydraulic liquid may cause difficulties if the air or gas forms bubbles and collapses as the fluid pressure fluctuates.

REVIEW QUESTIONS

1. Is the common bicycle pump classified as a fixed-displacement machine or a variable-displacement machine?

2. List several advantages or desirable features of a fixed-displacement pump.

3. List several advantages or desirable features of a variable-displacement fluid motor.

4. For what applications can reciprocating fluid motors be used?

5. For what applications can rotary fluid motors be used?

6. What is the chief difference between a compressor and a pump?

7. How can a pump be protected from overloading?

CHAPTER 4

Hydraulic and Pneumatic Symbols

A study of the operation of a hydraulic or pneumatic system is more convenient if a diagram of the system or circuit is available. The components in an installation and their connections can be determined from such a diagram. A *schematic* diagram indicates the functions of the various parts. In a given schematic diagram, for example, certain symbols and lines are used to represent a compressor, a cylinder, and a pipe from the compressor to the cylinder. The function of an air compressor is to increase the air pressure, the function of the pipe line is to transport air, and the function of the cylinder is to provide a means for doing work on a device or load. Thus, the schematic diagram shows the components and the connections between the components. A schematic diagram also indicates the operation of a system to a person who understands the symbols.

A schematic diagram of a hydraulic system or a pneumatic system is similar to a geographical road map. The symbols or language of the road map must be learned before the road map can be understood. Similarly, the symbols or language of a schematic diagram must be learned before the diagram can be used to trace a hydraulic or pneumatic system.

ANS GRAPHIC SYMBOLS FOR HYDRAULIC AND PNEUMATIC COMPONENTS

In the past, many different diagrams and symbols have been used—a practice which proved to be inconvenient and troublesome. A real need arose for a standard set of symbols. Accordingly, a number of conferences were held for the purpose of establishing a set of standard symbols for industrial hydraulic and pneumatic equipment.

A set of fluid power symbols approved by the *American National Standards Institute, Inc.* is shown in this chapter. These symbols can be of inestimable value to designers, installation and maintenance personnel, sales engineers, etc.

RULES FOR SYMBOLS

Symbols are used to show connections, flow paths, and functions of the components represented. Conditions occurring during transition from one flow-path arrangement to another can be indicated by symbols.

The locations of ports, direction of shifting of spools, and the positions of the control elements on the actual component are not indicated by symbols. Symbols are not used to indicate construction, and they are not used to indicate values, such as pressure, flow rate, and other component settings. The symbols can be rotated or reversed without affecting their meaning—except in lines to reservoirs, a vented manifold, and an accumulator.

ANS GRAPHIC SYMBOLS

General Symbols

A *working line* or *pipe line* carrying either oil or air under pressure.

A *pilot line*, usually running from a pilot valve to a master valve.

Flexible line, usually a rubber hose, such as used to connect a pivot-mounted cylinder.

Connector, as used in a piping layout.

Indicates the *flow direction* of the oil.

Indicates the *flow direction* of the air.

Indicates that a *line passes another line*, but is *not connected* to it.

A *pressure line* and a *connector*, representing a T-connection.

A *junction* of four pressure lines.

Elbow. These are usually not shown on a schematic diagram.

A *plug* or a plugged connection.

A pressure line with a *plugged connection*.

61

Lines for Hydraulic Equipment

------------------ *Drain line.*

Vented reservoir or tank for holding or storing oil.

A line to the reservoir, entering the reservoir *above the fluid level* in the reservoir or tank.

A line to the reservoir, entering the reservoir *below the fluid level* in the reservoir or tank.

Vented manifold.

Line with *fixed restriction.*

Lines for Pneumatic Equipment

------------------ *Exhaust lines* used to pipe the exhaust in from a control valve.

A *restricted pressure line*. Can be accomplished either by reducing the pipe size or by adding a restriction to the line.

Line for Both Hydraulic and Pneumatic Equipment

A *testing station* for a gauge connection. This symbol is a *combination of an exhaust or drain line and a plug.* A pressure gauge can be attached here.

Quick disconnect, without checks.

Quick disconnect, with checks.

Hydraulic Pumps

A *fixed-displacement pump* has a fixed number of gallons of oil per minute leaving the pump.

A *variable-displacement pump* is one in which the rate of oil flow leaving the pump may be varied.

Air Compressors

An *air compressor* with a *fixed displacement.* A fixed number of cubic feet of air per minute can enter the compressor.

Motors and Cylinders (Hydraulic and Pneumatic)

OIL   AIR

Rotary motor with a *fixed displacement.* The output shaft from the motor rotates, and the volume rate of flow through the motor is fixed.

63

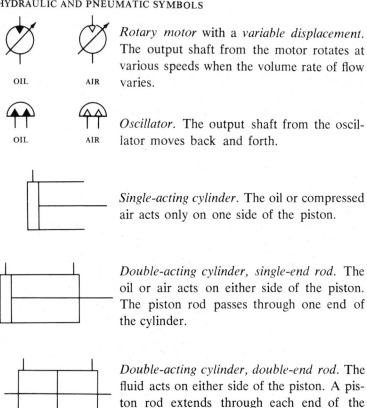

OIL AIR

Rotary motor with a *variable displacement.* The output shaft from the motor rotates at various speeds when the volume rate of flow varies.

OIL AIR

Oscillator. The output shaft from the oscillator moves back and forth.

Single-acting cylinder. The oil or compressed air acts only on one side of the piston.

Double-acting cylinder, single-end rod. The oil or air acts on either side of the piston. The piston rod passes through one end of the cylinder.

Double-acting cylinder, double-end rod. The fluid acts on either side of the piston. A piston rod extends through each end of the cylinder.

Motors and Cylinders (Pneumatic)

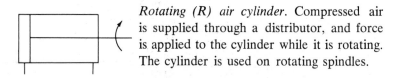

Rotating (R) air cylinder. Compressed air is supplied through a distributor, and force is applied to the cylinder while it is rotating. The cylinder is used on rotating spindles.

Miscellaneous Units (Hydraulic and Pneumatic)

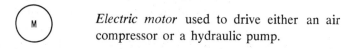

M

Electric motor used to drive either an air compressor or a hydraulic pump.

64

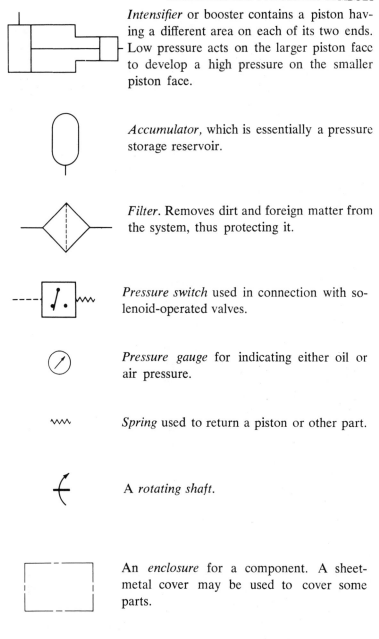

Intensifier or booster contains a piston having a different area on each of its two ends. Low pressure acts on the larger piston face to develop a high pressure on the smaller piston face.

Accumulator, which is essentially a pressure storage reservoir.

Filter. Removes dirt and foreign matter from the system, thus protecting it.

Pressure switch used in connection with solenoid-operated valves.

Pressure gauge for indicating either oil or air pressure.

Spring used to return a piston or other part.

A *rotating shaft.*

An *enclosure* for a component. A sheet-metal cover may be used to cover some parts.

Miscellaneous Units (Hydraulic)

Heat exchanger, used to cool the oil in a system. The radiator on an automobile is an example of a heat exchanger for cooling the circulating water.

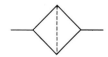

Strainer, which may include a screen for removing foreign particles.

Miscellaneous Units (Pneumatic)

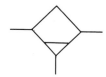

Separator. A device for removing water from a compressed-air system.

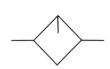

Lubricator. A device for controlling the amount of oil fed into a system. Lubrication is necessary to keep friction and wear to a minimum.

Storage tank or surge tank. The size of the tank depends on the requirements of the system.

Exhaust muffler. A device for reducing noise, as on an automobile.

Valves (Hydraulic and Pneumatic)

Check valve, which allows fluid to flow in only one direction.

Variable-restriction valve. The faucet in the home is an example of a valve that offers variable restriction to the flow of fluid.

66

Basic symbol for a *valve*.

An arrow inside the box indicates *internal flow*.

Examples of Hydraulic and Pneumatic Valves

Manual shutoff valve operated by either hand or foot.

Pilot-operated relief valve.

Shutoff valve having *two positions* of the valve and *two connections*.

Directional valve with *two positions* and *three connections*. This valve is used to control the flow and to give the desired motion and reversal of motion to the actuating piston or motor.

Directional valve with *two positions* and *four connections*.

Directional valve with *three positions, four connections,* and an arrangement that *closes* the valve when in the *center position*.

67

Examples of Hydraulic Valves

Sequence valve.

Pressure-reducing valve. The regulator provides the proper operating pressure for a system.

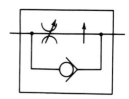

Pressure-compensated adjustable flow-control valve with bypass.

Examples of Pneumatic Valves

Sequence valve, sometimes called *resistance valve,* used to control the sequence of operation of air motion devices in a system.

Pressure-regulator valve, sometimes called a *reducing valve,* used to provide the proper operating pressure for a system.

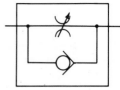

Speed-control valve with integral check.

Methods of Operation (Hydraulic and Pneumatic)

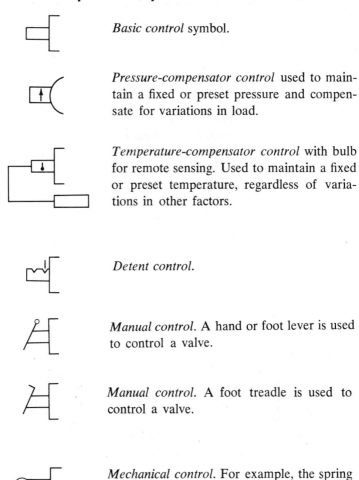

Basic control symbol.

Pressure-compensator control used to maintain a fixed or preset pressure and compensate for variations in load.

Temperature-compensator control with bulb for remote sensing. Used to maintain a fixed or preset temperature, regardless of variations in other factors.

Detent control.

Manual control. A hand or foot lever is used to control a valve.

Manual control. A foot treadle is used to control a valve.

Mechanical control. For example, the spring in a watch.

Electric motor control. This may be some type of switch.

Hydraulic pilot control, such as an oil pilot arrangement.

Air pilot control.

Servo control. A servo control compares two signals, and then exerts a control which is dependent on the difference between the two signals.

Solenoid control used to either open or close a valve by means of electric current.

Solenoid control operated by an *air pilot.* The air pilot system actuates the solenoid which, in turn, actuates the valve.

Solenoid control operated by a *hydraulic pilot.* The hydraulic system actuates the solenoid.

Thermal control.

Composite Actuators (And, Or, And/Or)

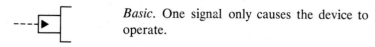

Basic. One signal only causes the device to operate.

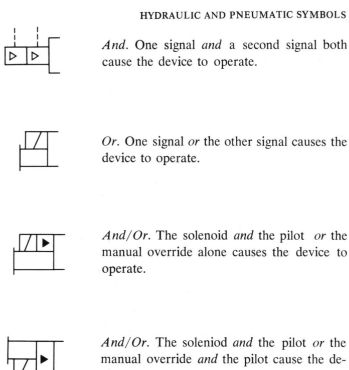

And. One signal *and* a second signal both cause the device to operate.

Or. One signal *or* the other signal causes the device to operate.

And/Or. The solenoid *and* the pilot *or* the manual override alone causes the device to operate.

And/Or. The soleniod *and* the pilot *or* the manual override *and* the pilot cause the device to operate.

And/Or. The solenoid *and* the pilot, *or* the manual override *and* the pilot, *or* the manual override alone causes the device to operate.

COMPOSITE SYMBOLS

The various hydraulic and pneumatic symbols can be combined, and a component enclosure symbol can be used to surround a symbol, or group of symbols, to represent an assembly. Examples of composite symbols are illustrated in Figs. 1, 2, 3, 4, and 5.

The *ANS* Graphic hydraulic and pneumatic symbols are often used in the study of the various aspects of hydraulic and pneumatic

71

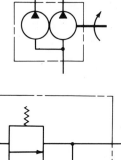

Fig. 1. A double pump with fixed-displacement, one inlet, and separate outlets.

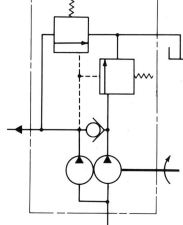

Fig. 2. A double pump with integral check, unloading, and relief valves.

devices in conjunction with hydraulic and pneumatic schematic diagrams. Occasionally, designations other than *ANS* Graphic symbols are used. The block-type symbols are sometimes used, as shown in Fig. 6. Since it is difficult to visualize the action within the component, the block-type symbols can be rather difficult to understand. The same circuit is shown with *ANS* Graphic symbols in Fig. 7. The *cross-sectional* symbols (Fig. 8) are actually a small-scale cross section of the component. These are relatively easy to understand, but considerable time is required to draw a complicated circuit when these symbols are used.

A four-way hand-operated control valve is illustrated in Fig. 9. This symbol has one disadvantage: a person attempting to check out the system cannot ascertain whether the valve is a two- or three-position valve, and he cannot have any perception of the configuration of the spool. Even though the valve is labeled with

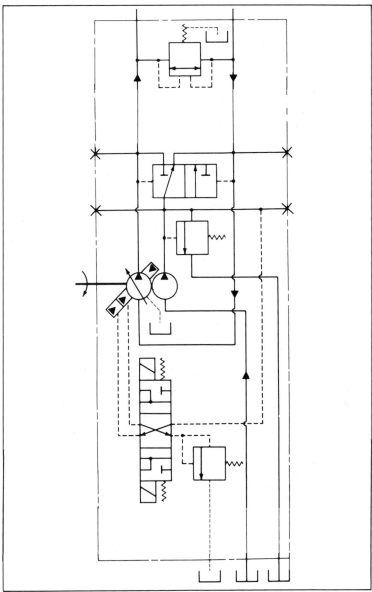

Fig. 3. Diagram of a pump having variable-displacement with integral replenishing pump and control valves.

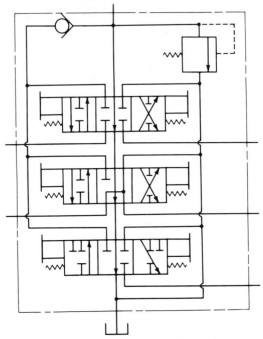

Fig. 4. A multiple, three-position, manual directional control with integral check and relief valves.

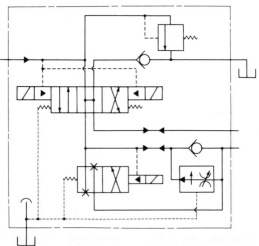

Fig. 5. Panel-mounted separate units furnished as a package (relief, two four-way, two check, and flow-rate control valves).

74

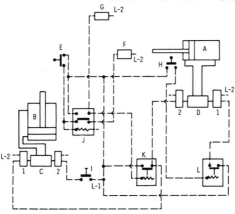

Fig. 6. A schematic diagram using block-type symbols in a pneumatic circuit.

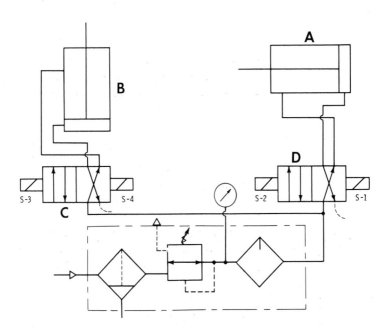

Fig. 7. A schematic diagram using USASI symbols in the same pneumatic circuit that is shown in Fig. 6.

75

the manufacturer's model number, the manufacturer's catalog is not always readily available when a failure occurs and repair is needed.

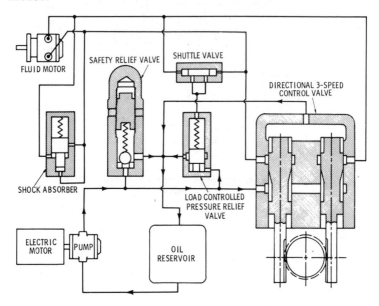

Fig. 8. Diagram of a fluid circuit in which the components are shown in cross section.

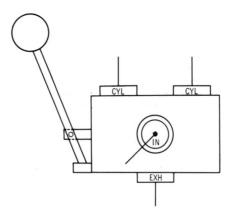

Fig. 9. Symbol indicating a four-way valve.

REVIEW QUESTIONS

1. Using *ANS* Graphic Symbols, draw a length of pipe showing: (a) A tee connection at one end and a cross on the opposite end of the pipe; and (b) A length of hose connected to a length of pipe, with a fixed orifice on the end of the pipe.

2. Using *ANS* Graphic Symbols, sketch a diagram of a hydraulic pipe line with the following components:
   a. Fixed-restriction.
   b. Variable-restriction.
   c. Pressure gauge.
   d. Relief valve.
   e. Check valve.

3. Using *ANS* Graphic Symbols, sketch a diagram showing a pneumatic filter, regulator, and lubricator in an enclosure. From the outlet line in the enclosure, show a tee line in which one branch of the tee has a regulator and pressure gauge, and the other branch has a shutoff valve.

4. Draw Fig. 1 in Chapter 1, using *ANS* Graphic Symbols.

5. Draw Fig. 2 in Chapter 1, using *ANS* Graphic Symbols.

CHAPTER 5

Pressure Boosters

A pressure booster is a device that can be used to increase fluid pressure; it can be activated by pneumatic or hydraulic pressure. The fluid that is intensified may be air, oil, water, or some other type of fluid. Boosters are often referred to as intensifiers. Basically, a booster is a cylinder that is equipped with a displacement chamber.

Three types of boosters that are now in common use are:

1. Pressure applied in one direction, with either a spring or gravity return in the other direction; intensified pressure is in one direction.
2. Pressure applied in both directions; intensified pressure is in one direction.
3. Pressure applied in both directions; intensified pressure is in both directions.

Boosters are used on the fluid systems of many industrial machines. It is often necessary to intensify the pressure of these systems.

TYPES OF PRESSURE BOOSTERS

Pressure boosters have a number of applications. They are used on pressure-burst testing machines, high-pressure clamping devices, molding machines, curing presses, spot-welding machines, riveting devices, marking machines, and many other machines.

Booster With Pressure Applied in One Direction

A booster that is actuated by hydraulic pressure and used to intensify the pressure of the oil in the intensifier chamber is diagrammed in Fig. 1. The formula for finding intensified pressure is:

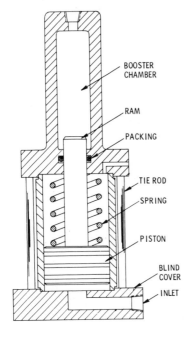

BOOSTER
CHAMBER

RAM

PACKING

TIE ROD

SPRING

PISTON

BLIND
COVER

INLET

Fig. 1. Diagram of a typical single-acting hydraulic booster.

$$P_1 A_1 = P_2 A_2$$

in which;

P_1 is the pressure in *psi* applied to the cylinder piston.
A_1 is the area in sq. in. of the cylinder piston.
P_2 is the theoretical pressure in *psi* output of the intensifier.
A_2 is the area in sq. in. of the intensifier ram.

Example: Calculate the intensified pressure if 1000 *psi* are applied to a cylinder piston having an area of 50.27 sq. in. (8-inch diameter piston), and the area of the intensifier ram is 3.142 sq. in. (2-inch diameter ram).

Substituting in the formula $P_1A_1 = P_2A_2$

$$1000 \times 50.27 = P_2 \times 3.142$$

$$P_2 = \frac{50.27 \times 1000}{3.142} = 15,999 \ psi$$

Thus an intensified pressure of 15,999 *psi* is produced from an applied pressure of 1000 *psi*—or an intensification of approximately 16 to 1. The intensifier shown in Fig. 1 has a spring-return feature, which reduces the applied pressure.

The use of an intensifier in a hydraulic system is illustrated in Fig. 2. When a single-acting booster is used, it can be actuated by a three-way directional control valve. A makeup tank should be used in conjunction with the intensifier.

If the parts which make up the booster (see Fig. 1) are studied, it can be seen that the booster resembles a hydraulic cylinder,

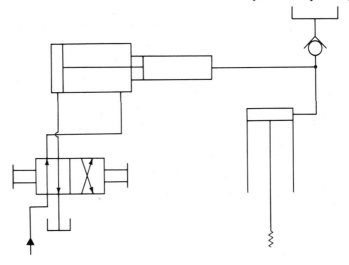

Fig. 2. Illustrating the use of an intensifier in a hydraulic system.

except for the booster chamber. Boosters can have mountings that are similar to those of nonrotating cylinders. It should be remembered that when a single-acting booster is used, the work that is required must be accomplished before the booster reaches the end of the stroke, or the pressure drops off immediately.

An air-operated single-acting booster that is used to intensify water pressure is shown in Fig. 3. The end of the booster in which the air pressure is applied closely resembles a pneumatic cylinder; the high-pressure end of the cylinder must be constructed of a noncorrosive material that is suitable for water under high pressure. To avoid rust, the ram should be made of stainless steel.

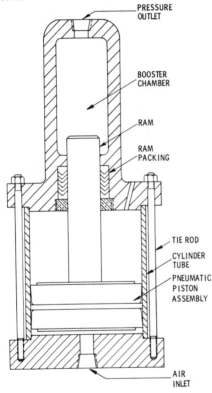

PRESSURE
OUTLET

BOOSTER
CHAMBER

RAM

RAM
PACKING

TIE ROD

CYLINDER
TUBE

PNEUMATIC
PISTON
ASSEMBLY

AIR
INLET

Fig. 3. Illustrating an air-operated single-acting pressure booster.

An air-operated hydraulic booster that is commonly used in industrial applications is shown in Fig. 4. High-pressure oil is emitted at the upper end of the booster.

An application of a pneumatic-water booster that is used to expand tubing for heating elements is illustrated in Fig. 5. This process eliminates the need for large expensive expanding-type mandrels.

Pressure Applied in Both Directions

This type of booster is similar to the single-acting booster, except for the piston being returned under pressure. Since full pressure is available on the booster stroke, this is an advantage over the spring-return type of booster. A four-way control, which can be manually, mechanically, or electrically operated, is used to actuate this type of booster. These boosters can have an intensification of 50 to 1—or even higher—but the seal around the booster ram can become a problem, because extremely high-pressure seals are not readily available.

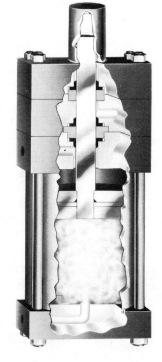

Fig. 4. An air-operated hydraulic booster. High-pressure oil is emitted from the upper end of the pressure booster.

*Courtesy Miller Fluid Power
Div., Flick-Reedy Corp.*

83

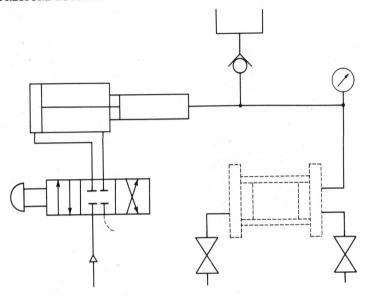

Fig. 5. Illustrating application of a pneumatic water pressure booster.

Pressure Applied and Intensified in Both Directions

A compact double-acting pressure booster which intensifies pressure in both directions of its piston movement is shown in Fig. 6. This unit is equipped with a piston that reciprocates automatically. Large areas of the piston are exposed to the force of the low-pressure pump. The fluid pressure of the low-pressure pump moves the piston, and causes the small-area pistons located at either end of the booster to force the oil outward and into the system under high pressure. This type of booster may be used with either a constant- or a variable-volume pump. In studying the parts that make up the pressure booster, note that the unit is equipped with a four-way control valve.

ADVANTAGES OF PRESSURE BOOSTERS

As aforementioned, pressure boosters have a number of applications on various industrial machines. Some of the advantages of boosters in a system are:

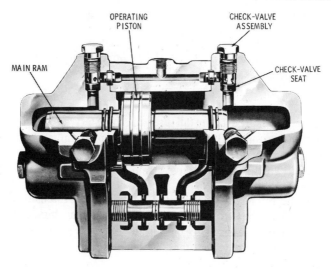

Courtesy Racine Hydraulics and Machinery, Inc.

Fig. 6. Cutaway view of a compact double-acting pressure booster.

1. Expensive high-pressure hydraulic pumps can be eliminated. If a booster is used, a pump can operate on only a few hundred pounds of pressure to produce an intensified pressure of several thousand pounds. If a pneumatic-actuated booster is used, only 80 to 90 *psi* are required.

2. Much higher pressure can be obtained by boosters than can be obtained by the high-pressure pumps.

3. Heat production is reduced. High-pressure pumps spill oil through the relief valve, causing considerable heat. This is eliminated when a booster is used.

4. Air-operated boosters may be used near explosive atmospheres, which permits the use of hydraulic fluid at high pressure. This is a distinct advantage, because the hydraulic power device can be eliminated.

5. Pressure boosters provide both high- and low-pressure fluid for a single pumping unit.

6. Low-pressure valves can be used in the system, because high pressure is normally developed beyond the control valves.

BOOSTER INSTALLATION AND MAINTENANCE

Pressure boosters should be installed and maintained properly for them to function correctly in their applications on the various industrial machines.

Installation

The methods used to install pressure boosters and nonrotating cylinders are similar. They should be mounted on a solid base. Those boosters with feet provided for mounting purposes should be mounted so that no excessive strain is placed on the mounting feet.

When the piping is installed, the piping on the intensified-pressure end should be strong enough to withstand the high pressure. Care should be exercised to be certain that the pipe and fittings are turned in securely and that no leaks occur around the pipe openings.

Pressure boosters should be installed in places where they are readily available for servicing. In installations that use water boosters, noncorrosive piping and fittings should be used on the intensified-pressure end wherever possible. Also, a pressure gauge of sufficient range should be installed in conjunction with the intensified-pressure end. Since there is considerable packing in pressure boosters, they should be kept away from direct hot blasts of air.

Causes of Failure of Boosters

Failures of pressure boosters may be attributed to the same problems that cause failures in nonrotating cylinders, as follows:

1. Dirt in the system will cause scoring of the cylinder walls, pistons, rams, and valve spools.
2. Heat may cause deterioration of the packing and seizure of close-fitting metal parts.
3. Lack of lubrication on the pneumatic-actuated pressure booster is a cause of seal and packing failures.
4. Broken springs on single-acting spring-return types of boosters cause the piston to fail to return to its starting position.
5. Misapplication is the cause of many fluid power component failures.

6. Packing failures can be due to wear or to excessive operating pressure.
7. If the check valve fails, all the intensified pressure flows back to the reservoir.

Maintenance

When a booster is serviced, it should be carefully dismantled and all parts cleaned with a good solvent. All residue should be brushed and soaked off the parts. All parts should be checked for wear, and any parts that show excessive wear should be either plated or replaced. Seals and packing should be replaced if they show any signs of wear. Any scores in the tubing should be cleaned up carefully.

All connections should be plugged if the booster is to be placed in stock. A pressure booster should be stored in a cool, clean, dry place.

After a booster has been repaired, it can be checked for correct input and output pressures by placing a pressure gauge on each end of the booster.

Pressure Produced By Intensifiers

Driving Cylinder		Intensifier Ram		Intensifier Ratio	Theoretical Intensified Hydraulic Pressure (P₂) When Using Input Pressure (P₁) Of		
Bore (in.)	Area (A₁) (sq. in.)	Dia. (in.)	Area (A₂) (sq. in.)		100 (psi)	500 (psi)	1,000 (psi)
4	12.566	1	0.785	16.00	1,600	8,000	16,000
5	19.635	1⅜	1.485	13.22	1,322	6,610	13,220
6	28.274	2	3.142	9.00	900	4,500	9,000
8	50.265	1	0.785	64.03	6,403	32,015	64,030
10	78.540	2	3.142	25.00	2,500	12,500	25,000
12	113.100	2	3.142	36.00	3,600	18,000	36,000
14	153.940	2	3.142	48.99	4,899	24,495	48,990

REVIEW QUESTIONS

1. What is meant by a pressure booster?

2. List the advantages of a pressure booster?

3. What are possible causes of failure of pressure boosters?

4. If a hydraulic pressure booster cylinder is 4 inches in diameter and is operated at 1500 psi, what is the intensified pressure if the booster ram is 1 inch in diameter?

5. In the preceding question, what is the volume of the intensified fluid if the length of stroke of the booster is twelve inches?

6. If the intensified pressure is 25,000 psi and the area of the booster ram is 1.25 square inches, what diameter is required on the cylinder end if the cylinder-end operating pressure is 750 psi?

7. An air-operated booster for water service has a cylinder with a 12-inch bore by 20-inch stroke. How much intensified fluid can be delivered per stroke if the ram diameter is 1.5 inches?

8. In the preceding problem, what is the intensified pressure if the cylinder operating pressure is 100 psi?

9. What is the advantage of using water, rather than oil, as the intensified medium in testing pipe?

10. What is the advantage of using an intensified fluid, rather than a mechanical device, to expand lengths of tubing that have already been bent into either a "U"-shape or a semi-circular shape?

11. Make a rough sketch of the following:
    (a) Single-acting booster actuated by air for intensifying water pressure. Piston to be returned by spring pressure.
    (b) Double-acting booster, intensification in one direction, actuated by oil for intensifying oil pressure

12. By means of a sketch, show how a booster should be connected to a hydraulic cylinder.

CHAPTER 6

Air Compressors and Accessories

The air compressor is a basic component of any pneumatic system (see Chapter 1). Air is compressed in a pneumatic system, so that it can be used to push, to pull, to do work, or to develop power. As air from the atmosphere enters the compressor, it is compressed by the machine to a higher pressure, and then discharged into a piping system. The compressed air may be used to drive air motors, air hammers, or other air devices.

Operation and installation of air compressors and their accessories are discussed in this chapter. The selection of pneumatic system components is also discussed here.

GENERAL TYPES OF COMPRESSORS

Two main types of air compressors are used in industry. Depending on the fluid action, they are classified as: (1) the *positive-displacement*, or pressure, type; and (2) the *velocity*, or dynamic, type.

Positive-Displacement, or Pressure, Type

In the pressure type of air compressor, the characteristic action is a volumetric or displacement action. Fluid pressure is developed primarily by a displacement action. The general construction of positive-displacement compressors may be divided into two groups, depending on the motion of the mechanical parts as: (a) *reciprocating;* and (b) *rotary.*

Reciprocating Compressors—The bicycle pump is a simple example of a reciprocating compressor (Fig. 1). As illustrated in the diagram, this simple machine includes a cylinder, piston, leather

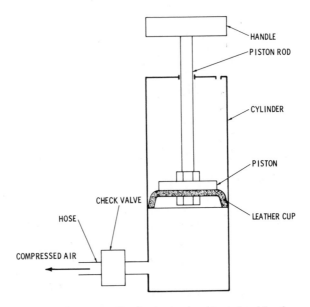

Fig. 1. Diagram illustrating the basic construction of a bicycle pump, as an example of a reciprocating air compressor.

cup, piston rod, handle, and check valve. If the piston is at the top of its stroke, the atmospheric air in the cylinder is compressed as the handle is pushed downward. When the air in the cylinder reaches a pressure slightly higher than the pressure in the line connected to the pump, the check valve opens and allows the air to be discharged from the cylinder. When the piston reaches the bottom of the cylinder, the check valve closes. As the piston is pulled upward again to the top of the cylinder, the flexible leather cup allows atmospheric air to enter the cylinder. As the piston is pushed downward, the leather cup acts as a check valve, sealing the space between the cylinder and the piston. The up-and-down motion of the piston is a reciprocating motion. A *vacuum pump* is a compressor that operates with an intake pressure lower than atmospheric pressure and with a discharge pressure near atmospheric pressure or higher.

A typical reciprocating positive-displacement compressor is illustrated in Fig. 2. The crankshaft may be driven either by an electric motor or by an engine. The motion of the crankshaft is transmitted by the connecting rod to the crosshead, which moves

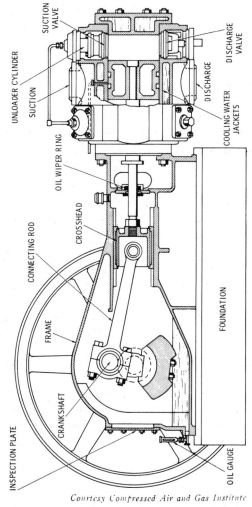

Fig. 2. Cutaway view of a typical reciprocating air compressor.

back and forth. The crosshead is also attached to one end of the piston rod, the other end being attached to the piston. Thus, the piston moves back and forth in the cylinder. The valves in the cylinder control the flow of air through the cylinder. The air is heated by heat of compression. As indicated in the diagram, water is circulated through water jackets surrounding the cylinder, which cools the air. The water jackets used on compressors are similar to those found on automobile engines. As shown in Fig. 2, the compressor rests on a base or foundation, and the crankcase requires oil for normal operation. Atmospheric air is admitted through the suction valve, and leaves the compressor through the discharge valve.

This machine (see Fig. 2) is called a "horizontal" compressor, because the piston moves in a horizontal direction. The term "single-stage" compressor indicates that the rise in pressure takes place only in a single cylinder. In a "two-stage" compressor, the air passes first through one cylinder and then through a second cylinder, a pressure rise occuring in each cylinder.

Various types of reciprocating compressors are in use. Two or more pistons may be driven by the same crankshaft, and the compressor may be driven by an electric motor, a steam engine, a diesel engine, an internal combustion engine, or a steam turbine. A direct-drive connection between the rotating shaft of the compressor and the rotating shaft of the drive unit is often used; however, belt and pulley drives are possible.

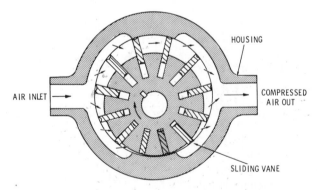

Fig. 3. Cutaway view of a sliding-vane type of rotary air compressor.

Rotary Compressor—The sliding-vane type of compressor (Fig. 3) is a rotary positive-displacement type of machine. The rotating member, with its sliding vanes, is set off center in the casing or housing. As the air enters, it is trapped between the vanes (which ride on the inside of the casing), and then carried to the discharge port.

Another type of rotary compressor uses lobes, rather than vanes, in the same manner (Fig. 4). The two lobes are mounted on paral-

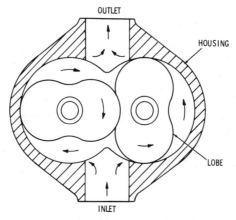

Fig. 4. Illustrating the lobe-type rotary air compressor.

lel shafts, and rotate in opposite directions. Air is drawn into the spaces between the lobes and the housing, and is moved from the inlet port to the outlet port. Timing gears located on one end of each parallel shaft maintain the proper relation between the lobes.

Velocity, or Dynamic, Type

In the velocity, or dynamic, type of compressor, the action between the air and a mechanical part involves an appreciable change in fluid velocity. The centrifugal compressor (Fig. 5) and the axial-flow compressor (Fig. 6) are examples of the dynamic type of compressor.

In the centrifugal compressor (see Fig. 5), there are four impellers; this machine is called a four-stage compressor. Each impeller is a circular member, or runner, with vanes. In each impel-

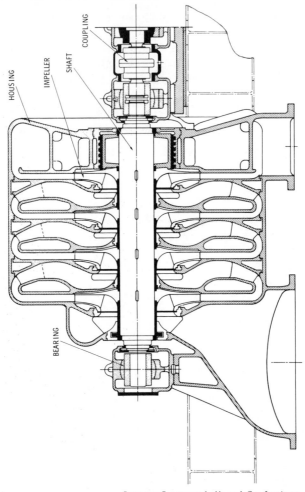

Courtesy Compressed Air and Gas Institute

Fig. 5. Cutaway view of a centrifugal air compressor.

ler, or stage, the pressure is boosted or increased a given amount. The total increase in air pressure is the result of the increases in pressure in the four separate stages. The high velocity of air from an impeller is reduced as the air pressure increases.

In actual operation, the air enters the first-stage impeller near the shaft, is thrown outward, and leaves the outer diameter of the

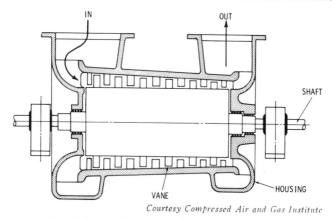

Courtesy Compressed Air and Gas Institute

Fig. 6. Illustrating an axial-flow air compressor.

impeller at a high velocity. The air leaving the first impeller then enters the second impeller, and the action is repeated, boosting the pressure in each stage.

In the axial-flow compressor (see Fig. 6), a rotating member with vanes is the basic component. Between each row of rotating vanes, stationary deflecting vanes are attached on the housing. The general motion of the air is parallel to the axis of the shaft, which explains the term "axial-flow" compressor.

COMPRESSOR LOCATION AND INSTALLATION

Various factors should be considered before installing a compressor. Considerable inconvenience and savings in maintenance costs may be avoided by carefully planning the location and installation of the compressor.

Location

The compressor should be located in an area that is easily accessible, but in an area that can be closed to personnel who are not connected with operation and maintenance of the compressor. The doors to the compressor room should be large enough for any part of the compressor to be moved out without dismantling any portion of the enclosure. The area that houses the compressor should be clean and well lighted. It should also be well ventilated, providing there is no excessive foreign matter in the atmosphere. When compressors are located in places such as grain elevators and flour mills, they should be protected from foreign material.

If several compressors are required, a problem can arise as to whether it is better to centralize all the compressors into a single area or to decentralize the areas, locating the individual compressors in separate areas. Generally, centralizing the compressors into a single area is more advantageous, because this type of location is more convenient to operate and maintain.

Installation

The foundation for a compressor should be solid. Many compressors weigh more than one ton, and the weight is usually very compact. The vibrating action of reciprocating compressors must be absorbed by the foundation. Frequently, compressor manufacturers furnish drawings which show the proper mass and base area characteristics of a good subsoil for the foundation.

Foundations made of concrete are definitely preferable to those made of other materials. The foundation should be constructed of a concrete mixture having a 1:2:5 ratio, which is one part of Portland cement, two parts of sand, and five parts of crushed stone. Another recommended mixture is one part of Portland cement to four parts of clean washed gravel. In the foundation, steel reinforcing rods should be placed lengthwise and crosswise to increase its strength The sand and cement should be mixed thoroughly before adding the gravel. Only enough water for a stiff plastic mixture should be added. The foundation should be allowed to set for at least a week before the compressor is installed, and the concrete should be kept moist during this period.

The foundation bolts should be located before pouring the concrete. When the forms for the foundation are made, a template for holding the foundation bolts in position may be made. The correct spacings for these bolts may be obtained from the drawing of the machine, or they may be measured directly from the bedplate of the compressor. The template may be made of boards, and the foundation bolts may be located on the boards, which helps to locate the bolts as the concrete is poured. Foundation bolts should not be cast solidly in the concrete. A steel sleeve consisting of about 3 or 4 inches of pipe may be slipped over each bolt. The steel sleeves remain in the foundation, and waste or similar material may be stuffed into each sleeve before pouring the concrete. The waste can be removed after the concrete has set.

The foundation should be placed well below the frost line—a depth of three feet or more may be required. A mat and piers are sometimes required for a solid foundation in locations where there is considerable dampness or a poor subsoil. If the mat and the foundation are not poured at the same time, vertical tie bars should be installed to tie the mat and the foundation together.

After the concrete has hardened, the bolt templates should be removed, and the machine prepared for mounting of the compressor. Wedges can be used to support the bedplate above the foundation temporarily. A dam high enough to hold the grout should be built around the foundation; it may be made of either wood or clay.

The grout should be prepared from a mixture containing one part of Portland cement to two parts of clean sharp sand, and mixed thoroughly with water to a consistency that can be poured and worked beneath the bedplate. As soon as the grout has set sufficiently, the dam and wedges should be removed. The remaining holes should be filled and the grout trimmed. After the grout has set thoroughly, the nuts on the foundation bolts should be tightened. The compressor should not be operated until the foundation is definitely and firmly set.

Air Intake—The air intake to the compressor is an important feature. If possible, the air should be brought in from outside the building. The intake opening should be located at sufficient distance above the ground level to prevent clogging the intake with leaves, rubbish, or waste paper. A distance of 8 or 10 feet above ground level is recommended, and the intake opening should be located in a shady or cool spot, if possible. Since moisture is extremely harmful to the pneumatic system, the intake opening should not be near a source of moisture.

If the intake opening of the compressor is located in a neighborhood where factories eject considerable impurities into the air, such as chemical fumes, flour dust, smoke, and particles of sand or cinders, a filter for the intake opening is extremely important.

The filter should keep out the impurities, and it should be equipped with louvers to keep out the rain.

Precautions should be taken to keep all the impurities out of the air intake lines to the compressor. If impurities are drawn into the compressor, they have a harmful effect on the valves, pistons, and cylinder walls of the compressor, and they also pass through the compressed-air lines to cause damage throughout the entire system.

The air intake lines should be made of materials that hold up well over a long period of time (Fig. 7). If pipe is used, it must

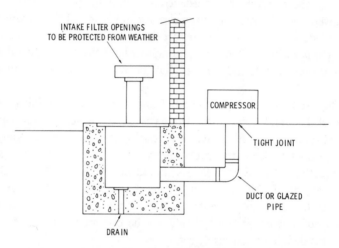

Fig. 7. Diagram showing the air intake inlet for an air compressor.

be free of scale and rust. The air intake may be made of poured concrete. If concrete is used, a solution should be painted on the surface of the concrete to prevent "powdering off," which may cause damage in the compressor.

The joint between the air intake line and the compressor should be airtight. This prevents picking up any dust that sometimes collects on the floor.

It is also important that the intake lines are sufficiently large in size. One recommendation is that the diameter of the intake line should exceed the diameter of the compressor inlet opening by 1 inch for each 10 feet of length.

Aftercooler—A system having only a receiver may be troubled with moisture in the distribution lines beyond the receiver (Fig. 8). One satisfactory method of preventing moisture difficulties is to remove the moisture immediately after compression.

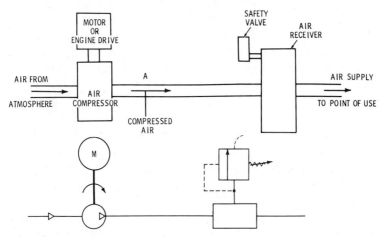

Fig. 8. Diagram illustrating a compressor and air receiver arrangement.

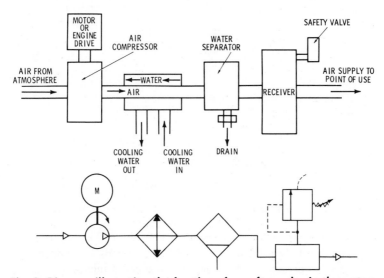

Fig. 9. Diagram illustrating the function of an aftercooler in the system.

99

An arrangement of a compressor, an aftercooler, a water separator, and a receiver is illustrated in Fig. 9. In the diagram, a cooling jacket surrounds the air pipe, and cooling water flows around the pipe to cool the compressed air to a temperature near that at the inlet. Thus the moisture in the compressed air is condensed, and then removed by the water separator.

Various types of aftercoolers are now in use. They are usually constructed of steel shells containing a nest of tubes. This type of aftercooler is illustrated in Fig. 10. A compressor with an aftercooler and a receiver is diagrammed in Fig. 11.

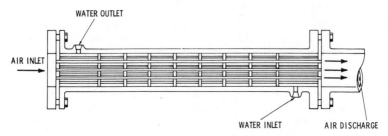

Fig. 10. Cutaway view of an aftercooler.

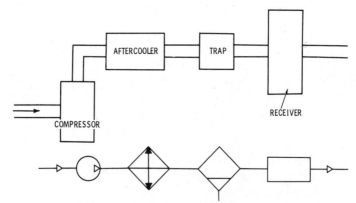

Fig. 11. Diagram illustrating a typical arrangement of the compressor, aftercooler, and receiver.

The aftercooler should be installed indoors to prevent freezing. Water is normally used for cooling. The aftercooler should be set near the compressor, and before the receiver, because any water that passes the separator can drop into the receiver.

After the compressor is installed, it is important to connect a proper distribution system from the compressor discharge outlet to the machine tools or cylinders where the work is performed.

Air Receivers—In Fig. 8, the air compressor is driven by either an electric motor or an engine. Air from the atmosphere enters the compressor, and then leaves the compressor at point A at high pressure. Reciprocating compressors deliver a pulsating flow—the air flow from a bicycle pump pulsates. When these pulsations occur only in a length of pipe, pressure waves can be built up—an action similar to the manner in which sound pressure waves are built up in the pipes of an organ. The pulsations may damage the compressed air system, and they may be noisy enough to disturb nearby workers. Also, the demand for compressed air may vary widely—from no demand to a very high demand, at times. If an air compressor delivers 100 cu. ft. of air per minute, for example, and a machine tool temporarily demands 120 cu. ft. of air per minute for proper operation, the air compressor is not capable of meeting the temporary demand for air. Hence, an "air receiver" or storage tank for compressed air is needed. In Fig. 8, the air receiver is placed in the piping from the discharge outlet of the air compressor. The air receiver dampens the pulsations created by the intermittent discharge of air by the compressor, and provides a capacity of reserve air for machine tool operations.

The air receiver or storage tank should be placed near the compressor if possible. The compressor is usually mounted directly above the storage tank of compressor units that are used to provide air for automobile tires in service stations. Reducing the length of the piping between the compressor and the receiver decreases pressure fluctuations in the air supply to the point of use.

A relatively large storage vessel for compressed air presents a safety problem. There is danger in the use of air receivers that are improperly constructed. In this country, the *A.S.M.E., The American Society of Mechanical Engineers,* has set up a code and standards for acceptable and approved construction of pressure vessels that are used for air receivers. Many states have passed laws that follow the *A.S.M.E.* code to ensure safe construction of these vessels. Only an approved air receiver or storage tank should be installed.

101

Most compressed air installations are concerned with compressed air that is used intermittently; periods of peak use may demand more air than the compressor can provide. The size of the receiver determines the peak demands that can be met. An undersized receiver limits the air supply; there can be no danger in using an oversized receiver. Large receivers, in general, improve the operation of the entire system.

Most air receivers are available with safety valves. The valve is set at a maximum safe pressure. If the storage pressure begins to exceed the safe maximum pressure, the safety valve opens and relieves the excess pressure. The *A.S.M.E.* has established a code for the rating of safety valves, and many states have incorporated the code in their laws that provide for the safe operation of air receivers. Receivers are also provided with pressure gauges, access doors, and supports for vertical mounting.

Moisture—Various experiments can prove that the normal atmospheric air is not completely dry. Atmospheric air contains some moisture. The moisture is in the form of transparent or invisible water vapor or steam. When atmospheric air is cooled to the so-called "dew point," the water vapor in the air is condensed to a noticeable liquid or drops of water.

The atmospheric air entering a compressor usually contains some water vapor. The air is then compressed, still carrying the water vapor. After it leaves the compressor, the compressed air containing the water vapor may experience various changes. The pressure may drop as the air passes through a machine tool, valve, or other component, and the temperature may drop. Thus, liquid water may be condensed. In some instances, pneumatic tools, for example, the temperature may drop low enough for the moisture to freeze and ice is formed.

The presence of water or moisture in a pneumatic system is undesirable. The condensed water moves along through the lines into the machine tools, valves, and other components. The lubricating oil may be washed away, and lack of proper lubricant may cause excessive wear, which may result in costly maintenance work. Freezing moisture may interfere with the proper operation of various tools and components; ice may clog the small openings in valves and ports.

Liquid or drops of water in the lines also may cause other difficulties. Drops of water may accumulate at low points in a line under particular conditions of flow. Drops of water may cause the line to be warm at a given time and cold at other times. During some conditions of flow, the drops of water bounce around inside the pipe, giving a "water hammer" effect, which may damage the pipe and fittings, in addition to creating a noisy disturbance. Intermittent operation of a pressure line, along with changes in temperature, may cause sufficient expansion (heating) and contraction (cooling) for the joints to leak.

Moisture is a real problem. It is impractical to eliminate water vapor before the air is compressed. Thus, steps must be taken to avoid difficulties that are due to condensation.

A typical arrangement of a compressor and receiver are shown in the diagram in Fig. 8. Moisture often condenses in the receiver.

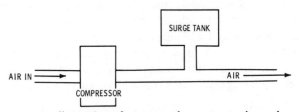

Fig. 12. Illustrating placement of a surge tank on the air discharge side of the compressor.

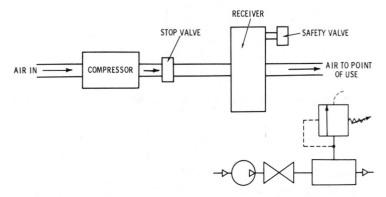

Fig. 13. Diagram illustrating an incorrect location for a stop valve.

Therefore, an adequate drain and valve are necessary at the bottom of the receiver to remove any liquid water that collects.

Air Discharge Pipe—The size of the pipe from the discharge outlet of the compressor should be no smaller than the pipe size of the compressor outlet. The discharge pipe leading to the receiver should be as short and direct as possible, to avoid pressure surges. Pressure surges occur frequently when long pipes are used. If pressure surges occur, they may sometimes be reduced by installing a tank or drum near the compressor discharge outlet, as illustrated in Fig. 12. The surge tank or drum acts as a cushion to prevent the transmission of pressure surges along the pipe.

Safety Valves—A safety valve on the air receiver protects the air distribution system. Care should be exercised to prevent any interference with the proper action of the safety valve.

An incorrectly located stop valve is shown in Fig. 13. The stop valve is placed between the air compressor and the safety valve. Proper action of the safety valve may be blocked by the stop valve. If the stop valve is closed while the compressor is operating, the compressor may build up a dangerously high pressure, and the safety valve cannot provide any protection; that is, it cannot act properly, because the pressure of the air at the compressor discharge outlet cannot reach the safety valve where it can be discharged safely.

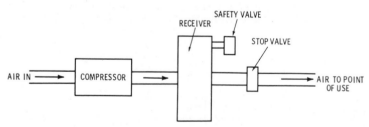

Fig. 14. Diagram illustrating correct location of a stop valve for safety in operation.

A safe location for the safety valve is shown in Fig. 14. In this instance, if the stop valve is closed and the compressor is operating, the safety valve can relieve the pressure if the pressure becomes dangerously high. Still another safe location for the safety valve is diagrammed in Fig. 15.

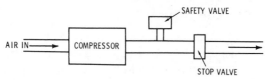

Fig. 15. Another diagram illustrating correct location of a stop valve for safety in operation.

OPERATION OF COMPRESSORS

In most instances, the recommendations of the manufacturers should be followed in compressor operation as well as in compressor installation. However, additional steps may be taken to provide safer, more economical, and more efficient operation of the equipment.

Cooling of Water

For proper operation, it is important that an adequate supply of cooling water is supplied for the aftercooler and the water jackets of the compressor. The cooling water should be turned on before the compressor is started. The importance of cooling water in the cooling system of the automobile engine is well known. Cooling water is of similar importance in a compressor unit.

Lubrication

Proper lubrication is essential for the proper operation of an automobile engine. In a similar manner, proper lubrication is important in the operation of an air compressor. Each air compressor requires a lubrication system. The compressor manufacturer provides instructions for the installation, operation, and care of the compressor. In these instructions, recommendations are made as to lubrication, and these recommendations should be followed. Lubricants should be purchased from dealers who guarantee the lubricant as satisfactory for the intended service. It is important to use clean oil of good quality. A compressor may require different grades of oil, such as one grade for the crankcase and bearings and another grade for the cylinder of the compressor.

105

Starting a New Compressor

A number of items should be checked before starting a new compressor which has been recently installed. Instructional booklets from the manufacturers should be consulted before attempting to start the new compressor. The recommendations of the manufacturer should be followed.

The air intake to the compressor should be checked to be sure that all parts are in place. The air intake should be cleaned of all foreign matter, such as dust and rust particles. Cotton waste should not be used for cleaning, to prevent waste particles getting into the intake air.

The crankcase should be carefully cleaned and filled to the recommended level with the proper grade of oil. All bearings, lubricators, and reservoirs should also be filled with the proper grade of oil. A check should be made to be certain that cylinder lubrication will begin as soon as the machine is started. This can be checked by first temporarily disconnecting the oil piping at the cylinder lubricator, operating the lubricator until oil issues forth, and then reconnecting the oil piping.

The compressor should be turned over several times by hand to make sure that all working parts are free. The cooling water should be turned on and checked to be sure that circulation is adequate.

The machine should be operated at first under no air load on the discharge side, and driving power should be applied only for short intervals. Checks should be made to see that: (1) there are no loose parts; (2) there is adequate lubrication; (3) the bearings are not overheating; (4) there is adequate cooling water; and (5) that various adjustments, such as the piston-rod packings, are correct. After a careful breaking-in period, normal operations can be begun.

COMPRESSED-AIR PLANTS

The chief purpose of a compressed-air plant is to provide compressed air at the proper pressure and quantity for the efficient operation of machine tools and other air machines. Care should be exercised in planning and selecting components for the compressed-air plant, thus obtaining the full benefits of compressed air and avoiding difficulties in operation of the components and the entire system.

Portable Air Compressors

A portable air compressor is a self-contained compressed-air plant that is mounted on a chassis for ready movement. The plant includes an air compressor, a driver for the air compressor, an air receiver, a cooling control, other control components, a lubricating system, a regulating system, and a starting system. Although portable air compressors are more frequently used on construction jobs, they may be very useful in indoor plant applications where portability is important.

Unit-Type Compressors

A unit-type air compressor or package-type air compressor is usually built complete with all the components needed to place the machine in operation. The compressor size is usually 20 horsepower, or smaller. The compressor, motor, and accessories are assembled on a base which, in turn, is mounted on an air storage tank or receiver. Most unit-type air compressors are air cooled. The unit-type compressors are commonly found in service stations, garages, paint shops, small machine shops, and similar plants.

Compressor Regulators or Controls

Usually, compressed air is required in varying quantities; a compressor regulator or control may be arranged to vary the air delivery of the compressor to satisfy the demand. A control system can be actuated by a pressure responsive device or pilot. If the air pressure becomes too high, the pilot acts to reduce or to interrupt air delivery. If the air pressure becomes too low, the pilot acts to restore or to increase the air delivery.

Various types of regulating devices may be used to control air delivery. In one type of regulating system, the compressor inlet valves are held open mechanically during the suction and compression strokes; hence, no air is compressed in the cylinder. In another type of system, a valve is used to close the intake line completely; this keeps air from entering the compressor. In another system (as with systems driven by steam engines, diesel engines, and gasoline engines), the speed of the compressor is varied, according to pressure changes sensed by the pilot.

Large compressors may be kept running while air is being delivered. Small compressors may be shut down with an automatic start-stop action; a pressure-actuated switch is used either to open or to close the motor circuit.

PLANNING A COMPRESSED-AIR PLANT

The compressed-air plant includes one or more compressors (including the motor or engine for driving), controls, air intake filter, aftercooler, air receiver, interconnecting piping with fittings and accessories, and a distribution system for delivering the compressed air to various points for use.

Before considering details of a plant (either a new plant or an expansion of the present plant), it is well to study the various specific plant requirements. The following features should be considered:

1. What are the present requirements, or needs, for compressed air, and what are the requirements of a possible future expansion?

2. What are the air requirements, both quantity and pressure, for the plant during normal operations? The length of time the tools and machines are operated, the quantity of air needed, and the pressure required should be considered. Perhaps, different pressure ranges are required, because some types of air tools require a pressure of 90 pounds per square inch gauge, whereas other groups require a different pressure. Stand-by needs and the possible effect of an inadequate supply of air should also be considered. Many of the air devices in a plant operate almost continuously; others operate intermittently, but require a large supply of air when operating.

3. What automatic protection should be provided for unattended operation?

4. What is the capacity of the present air supply equipment? Can the air supply be expanded, or is it being used to its fullest capacity?

5. The relation between operating cost and original cost should be considered.

Once the requirements for a compressed-air plant are established, various manufacturers can supply information on available equipment. For example, the standard sizes of compressors and their corresponding costs can be obtained from manufacturers and equipment suppliers.

COMPRESSOR SELECTION

Compressor selection presents a problem of considering the types of compressors available, the number of compressors needed, and the location of the compressors. Terms that are commonly used should be defined first.

Rated machine speed usually refers to the operation at the best or nominal compressor shaft speed. For example, in a motor-driven compressor, the "rated," "nominal," or most efficient speed of the compressor shaft is usually expressed as 250 *rpm* (revolutions per minute).

In reciprocating compressors, *piston displacement* is defined as the net volume actually displaced by the piston at a specified rated speed. Usually, piston displacement is given in *cfm* (cubic feet per minute).

Actual capacity is the quantity of gas or air actually compressed and delivered to the discharge system at a given rated speed of the machine under specific pressure conditions.

Free air is defined as air at atmospheric conditions at a specific location. Free air may be applied to displacement or capacity as a measure of volume. To be definite, the conditions for free air should be specified. Since altitude, barometric pressure, and air temperature may vary at different locations, the term "free air" does not indicate air under identical conditions.

Typical specifications for a given air compressor can read as follows: rated shaft speed, 280 *rpm;* piston displacement, 1550 *cfm;* and actual air delivery, 1330 *cfm*. These specifications indicate that the air compressor can deliver 1330 *cfm* of "free" air when the temperature and pressure of the surrounding air at the air intake are 60°F. and 14.7 *psi* absolute, respectively.

Volumetric efficiency is defined as the ratio of the actual capacity of the compressor divided by the piston displacement; this term is usually expressed in percent. In the preceding example, the volumetric efficiency is (1330 ÷ 1550) 100, or 86 percent.

For another typical example, the specifications of a commercially available two-stage reciprocating air compressor that is mounted on a storage tank and driven by an electric motor may read as follows:

> piston displacement = 67.5 *cfm*
> maximum discharge pressure = 175 *psig*
> rated shaft speed = 1450 *rpm*
> storage tank size = 80 gals.
> motor horsepower = 15 *hp*
> bore and stroke = 6 × 3½ ×2 ¾

The bore and stroke specification indicates that the first-stage piston diameter is 6 inches, the second-stage piston diameter is 3 1/2 inches, and the length of stroke for each stage is 2 3/4 inches.

The positive-displacement compressors are generally used where relatively low capacities and high air pressures are required. The dynamic type of air compressor is more satisfactory where relatively low air pressures and high capacities are needed for efficient operation.

Since the selection of a satisfactory compressor is largely determined by the requirements of pressure and capacity, primary consideration should be given the types of machines and machine

Table 1. Air Requirements For Pneumatic Tools

Inlet Pressure (psiq)	Tool	Air Rate (cfm Free Air)
90	Grinder, 6-inch, 8-inch wheels	50
90	Sander, rotary, 9-inch pad	53
90	Hammer, chipping, heavy	39
90	Plug drills	40-50
90	Drills, steel, up to 1¼ inch, weighing 30 lbs.	95
80	Sand blast, ½-inch diam. nozzle	340
80-90	Hoist, one-ton	1
80-90	Paint spray gun	8

tools that are to be connected to the air distribution line. If pneumatic tools that require a pressure of 90 *psig* at the tool inlet are to be used, a compressor that is capable of a discharge pressure higher than 90 *psig* must be selected to withstand the pressure drop in the system between the compressor discharge outlet and the tool inlet. The required capacity is determined by the demand of each pneumatic device that is to be connected to the air distribution line.

The requirements of the various pneumatic tools are listed in Table 1, as taken from the *Compressed Air and Gas Handbook*. For example, a 6-inch grinder (as listed) operating continuously requires an air flow of 50 *cfm* of free air. This type of data may be obtained from the manufacturers of pneumatic equipment, and may be used to study the air requirements for the specific pneumatic tools. A decision as to the number of compressors required involves the problems of cost, maintenance, efficiency, control, and stand-by capacity, which may vary for each plant.

REVIEW QUESTIONS

1. Why should the foundation for an air compressor be solid?

2. Why should the air intake be located in an area where the temperature is low and the moisture content of the air is low?

3. What is the purpose of an air receiver?

4. What is the function of an aftercooler?

5. What factors determine the location of a safety or relief valve in the system?

6. Why is proper lubrication important in an air compressor?

7. What is the difference between a single-stage and a two-stage air compressor?

8. Draw a schematic diagram of Fig. 12, using *ANS* Graphic Symbols.

9. Draw a schematic diagram of Fig. 14, using *ANS* Graphic Symbols.

10. Draw a schematic diagram of Fig. 15, using *ANS* Graphic Symbols.

CHAPTER 7

Hydraulic Power Devices

The functions of a hydraulic power device are to provide hydraulic fluid under pressure to a hydraulic system and to provide a place for storing the oil that is not in use. The amount of fluid delivered depends on the capacity of the pump. The capacity of the pump depends on the fluid displacement per revolution of the pump multiplied by the speed at which the pump is to be operated. Pump capacity is usually measured in gallons per minute.

The heart of the power device is the hydraulic pump. Other components of the power unit include the oil reservoir, intake filter, pressure gauge, pressure relief valve, coupling for connecting the pump and the driving means, which may be an electric motor, an internal combustion engine, or an air motor, and the internal piping. Many power devices are equipped with aftercoolers, heat exchangers, or heaters.

THE HYDRAULIC PUMP

The pump creates a partial vacuum (pressure below atmospheric) on the intake side as the internal mechanism starts through its cycle; then the atmospheric pressure acting on the oil in the reservoir forces the oil into the pump. The pump, as the cycle progresses, traps this oil, and forces it through the outlet under pressure.

It should be remembered that the distance the fluid can be raised vertically depends on the atmospheric pressure acting on the surface of the fluid and the amount of vacuum created within the pump. The atmospheric pressure varies with the altitude; at sea level it is commonly taken as 14.7 *psi*, and the theoretical maximum lift at this point is about 34 feet. The theoretical lift and the actual distance which a fluid can be raised may vary greatly, as mechanical imperfections in the pump, pipe friction, and wear of the pump parts should be taken into consideration. Although the pump can raise the fluid only a theoretical maximum of about 34 feet, it can force the liquid to much greater heights, depending on the force exerted on the fluid.

Generally, the requirement for raising the liquid into the pump amounts to a maximum distance of a few feet. Most storage tanks (or reservoirs, as they are commonly termed in hydraulics) have the pump mounted on top of the tank, and the oil has to be raised only a short distance.

Pumps may be divided into two general classifications—the constant- or positive-displacement type and the variable-displacement type. Since pumps are the heart of hydraulic systems, as much as possible should be learned about them.

Constant-delivery pumps have broad application in industry, and they are used in great numbers. Most constant-delivery pumps are of the rotary type, but some of the larger ones are of the reciprocating-piston type. The nomenclature for the constant-delivery pump is *pump—fixed delivery,* and for the variable-delivery pump it is *pump—variable-delivery* (see Chapter 4). Let us consider some of the various designs of hydraulic pumps.

(A)   (B)   (C)

Courtesy Roper Pump Company

Fig. 1. Illustrating the movement of a liquid through a gear-type hydraulic pump: (A) liquid entering the pump; (B) liquid being carried between the teeth of the gears; and (C) liquid being forced into the discharge line.

Gear-Type Pumps

These pumps may use pumping mechanisms that consist of: (1) internal gears, (2) external gears, or (3) a combination of internal and external gears. A gear pump is a rotary pump in which the gears rotate to cause the pumping action. Fig. 1 shows how the fluid moves through a gear-type pump. The liquid enters the pump (Fig. 1A), is carried between the teeth (Fig. 1B), and is forced into the discharge line (Fig. 1C). As the unmeshing action of the gears forms a vacuum, the atmospheric pressure acting on the surface of the liquid in the reservoir causes the liquid to fill up the tooth spaces. The liquid is then carried between the teeth and the pump case to the opposite side of the pump. As the gears mesh, the liquid is forced into the discharge line. This is a positive-displacement action.

It should be remembered that pumping action takes place very rapidly, as the shaft speed of the gear pump may be as high as 2700 *rpm*, although these speeds usually range from 750 to 1750 *rpm*. Gear pumps have been designed to operate up to 3000 *psi,* but the most popular pressure range is up to 1000 *psi.*

A cutaway section of a spur-gear type of pump is shown in Fig. 2. Gear-type pumps are also designed to use spiral, herringbone, and helical gears.

Fig. 2. Cutaway section of spur-gear type of hydraulic pump.

Courtesy Roper Pump Company

The names of the various parts of a typical spur-gear type of pump may be learned from the diagram in Fig. 3. Note the simplicity of the pump design.

115

Internal and external gear-shaped elements are combined in the *Gerotor* mechanism (Fig. 4). In the *Gerotor* mechanism, the tooth form of the inner *Gerotor* is generated from the tooth form of the outer *Gerotor*, so that each tooth of the inner *Gerotor* is in sliding contact with the outer *Gerotor* at all times, providing continuous fluid-tight engagement. As the teeth disengage, the space between them increases in size, creating a partial vacuum into

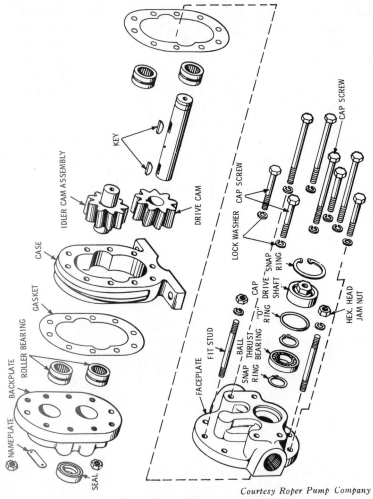

Courtesy Roper Pump Company

Fig. 3. Diagram and parts list for a gear-type pump.

which the liquid flows from the suction port to which the enlarging tooth space is exposed. When the chamber reaches its maximum volume, it is exposed to the discharge port; and as the chamber diminishes in size due to meshing of the teeth, the liquid is forced from the pump. The various parts of the *Gerotor* pump are shown in Fig. 5.

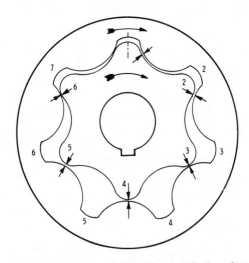

Courtesy Double A Products Co.

Fig. 4. Schematic diagram of the Gerotor mechanism.

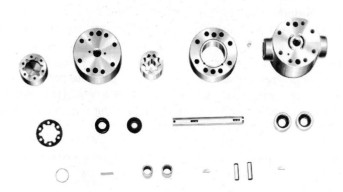

Courtesy Double A Products Co.

Fig. 5. Basic parts of a Gerotor pump.

Gears that make up the pumping elements for gear pumps are made of various materials, depending on the severity of the service. The gears may be made of bronze, cast iron, synthetic material, or hardened steel. In order to keep internal leakage to a minimum, tolerances between the gears and the pump housing or casing should be kept to a minimum.

Gear pumps have several advantages:

1. *Simple in design.* They are very compact and require little space for mounting. Their simplicity makes them suitable for rugged service and comparatively long life.
2. *Low in cost.* These pumps are produced in large quantities, and their cost is lower than many other types.
3. *Few moving parts.* This makes them easy to maintain.
4. *Quiet in operation.* Since the elements of these pumps run in oil, they are quiet in operation.
5. *Some gear pumps are reversible.* These are the spur gear and the helical gear types.

Vane-Type Pumps

The vane-type pump is also a rotary pump, and it also operates on the principle of increasing the size of the cavity to form a vacuum, which allows the fluid to fill up the space; the diminishing volume causes the fluid to be forced out of the pump under pressure. Vane-type pumps are either constant-delivery or variable-delivery in type. The vanes are flat rectangular segments of hardened steel that are held to very close tolerances. These vanes operate in a rotor which is slotted to accommodate the vanes. Some types of vane pumps use a single vane per slot—one type uses two vanes per slot. A section of a vane-type pump is shown in Fig. 6. Note that none of the moving parts contacts the pump housing. If the moving parts become worn, the whole pumping cartridge can be removed easily and replaced without disconnecting the piping, which saves time in maintenance and repair. Fig. 6 also shows the shaft bearings, fluid passages, and pump shaft.

A vane-type pump having two vanes per slot is shown in Fig. 7. Fig. 8 shows how this pump is hydraulically balanced in order

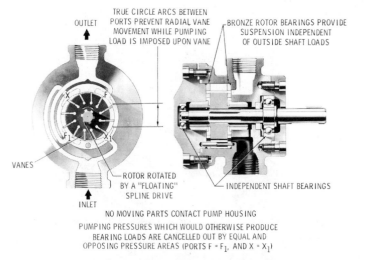

Fig. 6. Sectional view of a vane-type pump.

to eliminate radial bearing loads. The two diametrically opposed inlet sections are interconnected to provide equal flow and pressure to the pumping chambers. Also, the two discharge sections are diametrically opposed and are interconnected. On the rotor, the radial pressure load in one direction is always balanced by an equal pressure load in the opposite direction. By using a film of fluid, under pressure, between the shaft and the shaft bearing, the shaft floats in oil.

Axial thrusts are eliminated by equalized fluid forces which are maintained on both faces of the free-floating rotor. Opposing inlet sections and outlet sections are cross-connected through the cam ring. Forces excited by the oil pressure on one face of the rotor are balanced exactly by the forces acting on the opposite face. The symmetrical and balanced loading of the pumping chambers eliminates "cocking" of the rotor. This prevents wear of the port plates. The vanes are held against the cam ring by hydraulic force (Fig. 9).

Vane-type pumps of the constant-delivery type are available as follows:

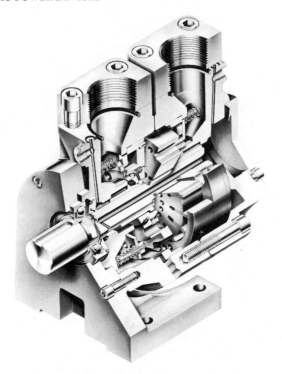

Fig. 7. A vane-type pump having two vanes per slot.

1. Single-stage pumps (see Figs. 6 and 7).
2. The two-stage type, which is two single-stage pumps hooked in series and placed in one body.
3. Double vane-type pumps, which are two pumps mounted end to end on a single shaft, in which one pump usually delivers more fluid than the other.
4. Combination pumps, which may be two pumps mounted on a common shaft, with the larger pump pumping at low pressure and the smaller pump delivering at high pressure. The high-pressure pump does not "cut in" until the load is met, and the low-pressure pump is unloaded automatically while the high-pressure pump delivers. During the low-pressure cycle, the high-pressure pump is pumping at

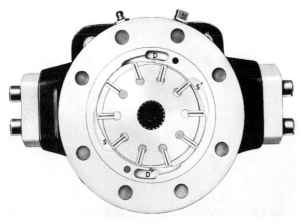

Courtesy Parker-Hannifin Corp.

Fig. 8. Showing how the vane-type pump is hydraulically balanced to eliminate radial bearing loads.

Courtesy Parker-Hannifin Corp.

Fig. 9. The vanes are held against the cam ring by hydraulic force.

low pressure. This type of pump is often called a "hi-low" pump.

A vane-type pump offering the variable-volume feature is shown in Fig. 10. Note the vane slots in the rotor. The rotor is made

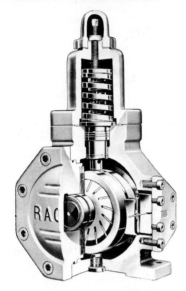

Fig. 10. Variable-volume vane-type pump.

of a heat-treated alloy steel. The pressure chamber ring is also made of a heat-treated alloy steel, and is hardened and ground. Vanes are made of a tool steel, and they are heat-treated and ground.

The bonnet at the top of the pump houses the pressure-compensating governor. This increases the versatility of the pump. The action of the pump may be altered by changing the governor units. In addition to the spring-type governor shown, other types may be used, such as the manual volume control, hydraulically operated two-pressure type of governor and the solenoid-operated two-pressure type of governor.

Rotary Piston-Type Pumps

Rotary piston pumps are either radial or axial in design. In the radial design, the pistons of the pump are arranged radially around a rotor hub. In the axial type, the pistons are arranged parallel to the shaft of the pump rotor.

Either the radial pump or the axial pump may be designed for either constant-displacement or variable-displacement. The variable-displacement types are more complicated in design, but they have the advantage of being able to change the discharge volume at a given speed. This cannot be done with a constant-displacement type of pump.

A plan view section of a constant-delivery radial piston-type pump is shown in Fig. 11. This pump is of the radial rolling-piston fixed-stroke type, and is used for discharge pressures up to 3000 *psi*. In this type of pump the volume of fluid discharge varies

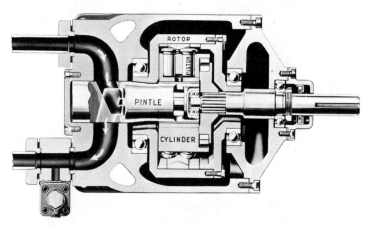

Fig. 11. Constant-delivery radial piston-type pump.

with the size of the unit and the speed at which the pump shaft is driven. In this design, the alloy-iron cylinder, which is lined with an antifriction metal, rotates on a fluid film about the stationary pintle. This pintle is pressed into the pump case. The ports and drilled passages in the pintle direct the fluid to and from the cylinder bores. The pistons are closely fitted to the cylinder bores. A single spot of the lapped beveled surface on the top of the piston contacts the hardened and ground beveled surface inside the thrust ring. Note that the balanced rotor and thrust-ring assembly is mounted to the left-hand side of the pintle center line and is free to rotate on antifriction bearings. Also note that the drive shaft is mounted on two antifriction bearings and directly connected to the cylinder through a splined floating coupling.

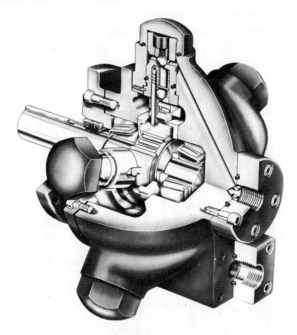

Fig. 12. A constant-displacement radial piston-type pump.

In operation, the driving force is applied to the input shaft which drives the cylinder, piston, and rotor assembly through a floating coupling. Centrifugal force acts on the pistons to force them against the thrust rings, and the rotor rotates with the cylinder. During the lower one-half revolution, the pistons move outward to pull fluid through the pintle to fill the cavities of the cylinder bores. During the upper one-half revolution, the pistons move inward to discharge the fluid from the cylinder cavities to the pintle and pressure line. With the contact spot offset from the axis of the piston and with the rotor eccentric from the cylinder, the pistons roll and reciprocate simultaneously. Both motions are uniformly accelerated and decelerated.

Fig. 12 shows a constant-displacement radial piston-type pump capable of producing pressure of 5000 *psi*. Note that it is designed for flange mounting for adaption to an electric motor drive.

A plan view section of a variable-delivery radial piston-type pump with a small built-in constant-delivery gear pump is shown in Fig. 13. In principle of operation this pump is similar to that shown in Fig. 11, except for the addition of the slide block. The slide block moves the rotor assembly and controls the stroke of the pistons. The slide block controls the pump delivery. In Fig. 13, the slide block is moved with a handwheel which actuates a screw. Other actuators are used, such as gear-head motors, hydraulic servo controls, and automatic unloading controls.

In Fig. 14, note that the slide block is at the right-hand side of the cylinder barrel center line. The reciprocating motion is imparted to the pistons, so that those passing over the lower port in the pintle are delivering oil to that port while those passing

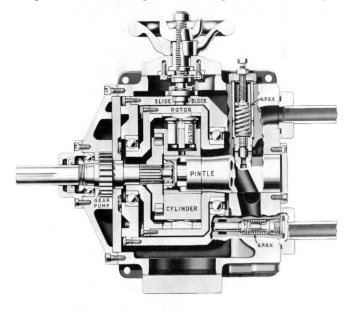

Fig. 13. A sectional view of a variable-displacement radial piston pump.

over the upper port are filling with oil. Since the piston and movement of the slide block can be controlled very accurately, the pump delivery can be accurately controlled from zero to maximum capacity of the pump. When the center lines of the cylinder and the rotor coincide, no reciprocating motion is imparted to the pistons as the unit rotates and there is no delivery of fluid from the pump.

The function of the small gear pump is to supercharge partially the main fluid power system and to actuate hydraulic controls. There are small radial piston pumps capable of producing pres-

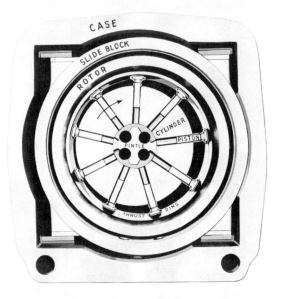

Courtesy The Oilgear Company

Fig. 14. Sectional view of a variable-delivery pump.

sures up to 10,000 *psi*. These usually produce only small volumes of fluid.

The axial piston-type pump may be of either the constant- or the variable-delivery type. These pumps work well in the high-pressure range up to 5000 *psi*.

In operation, the driving means of the pump rotates the cylinder barrel. The shoe retainer, spring-loaded toward the cam plate, causes the axial reciprocation of the pumping pistons that are confined in the cylinder. The angle of the cam plate limits the piston stroke and the quantity of oil delivered. In the variable-volume pump, a mechanism is used to change the angle of the cam plate. The volume may be changed from zero to the maximum output of the pump. The mechanism actuates a free-swinging hanger which is attached to the cam plate for changing the angle of the cam plate. This mechanism may be a handwheel control, a cylinder control, a pressure-compensating control, or a stem control.

In a constant-delivery axial piston-type pump the cam plate angle is fixed—see the exploded view (Fig. 15) and the cross-section-

Fig. 15. Exploded view of an axial piston-type pump.

al view of the constant-delivery pump (Fig. 16). Some of the advantages of an axial piston-type pump are: compact design, hydraulic balance, and satisfactory operation on a wide range of oil viscosities.

Reciprocating Piston Pumps

Reciprocating piston pumps which are used in industrial hydraulic applications are usually of large capacity. Some have capacities of several hundred gallons per minute. Reciprocating piston pumps are often used to supply fluid to a central hydraulic system. Hydraulic fluids handled by this type of pump are water, soluble oil in water, fire-resistant hydraulic fluid, or hydraulic oil.

A five-cylinder six-inch stroke pump with built-in gearing is shown in Fig. 17. Fig. 18 shows a five-inch stroke pump which has five cylinders. Note the crankshaft and connecting rods.

Reciprocating piston pumps may be designed with 3, 5, 7, or 9 pistons. The *Aldrich* direct-flow principle is illustrated in Fig.

127

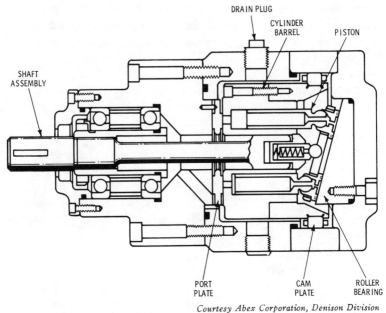

Courtesy Abex Corporation, Denison Division

Fig. 16. Cross-sectional view of a constant-displacement axial piston-type pump.

19, and the location of the parts is identified. Close-clearance design gives high volumetric efficiency in handling liquids in a wide range of viscosities. The suction and discharge manifolds are separate sections, which are bolted to a forged-steel working barrel. Removable stuffing boxes are provided to permit adapting a wide range of plunger sizes to the pump.

OIL RESERVOIR

The oil reservoir plays an important part in the hydraulic system. The *ANS* Graphic Symbol for a hydraulic reservoir is shown in Fig. 20. The oil reservoir has several functions:

1. It is a storage tank for the oil.
2. It provides a means of keeping the oil at the correct operating temperature by having a large enough quantity of oil in the reservoir to cool down the oil as it returns from

128

Courtesy The Aldrich Pump Company, Division of Ingersol-Rand Company

Fig. 17. A five-cylinder reciprocating piston-type pump.

the system or as it is expelled by the relief valve. The reservoir's shell allows the heat to radiate from the oil. Sometimes it may be necessary to give the reservoir some help by adding a cooling means. In very cold climates, it may be necessary to use an immersion-type heater to warm the oil to the correct temperatures. An immersion-type heater is submerged in the oil in the reservoir.

3. It provides a base for holding the components of the power device.

4. The reservoir, properly designed, slows up the oil flow as it returns from the system and keeps it from foaming. Liquids, under force, suddenly shot out against a solid surface may foam, because air mixes with the liquid.

5. Since the reservoir is located in the lowest spot in the system, it collects on the bottom of the reservoir the dirt and foreign particles that gather in a system; then they can be removed.

129

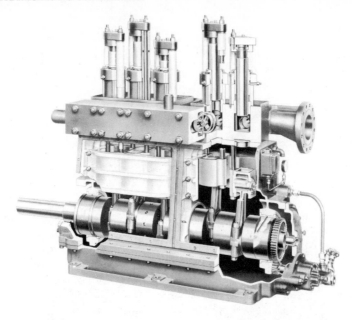

Courtesy The Aldrich Pump Company, Division of Ingersol-Rand Company

Fig. 18. Showing the internal parts of a reciprocating piston-type pump.

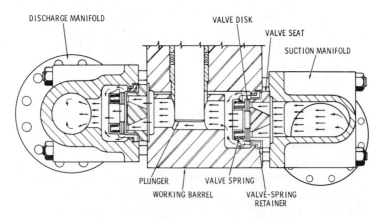

Courtesy The Aldrich Pump Company, Division of Ingersol-Rand Company

Fig. 19. Illustrating the direct-flow principle in the Aldrich reciprocating piston pump.

130

Fig. 20. The ANS Graphic Symbol for an oil reservoir.

The reservoir is equipped with baffles. Properly designed baffles keep the oil from foaming, and cause the sediment to drop to the bottom. If the bottom of the reservoir is made to slope, dirt will settle at one end. It is recommended that the bottom of the reservoir be kept off the floor, so that air can circulate underneath.

The reservoir is usually made of plate steel of sufficient strength to furnish ample support to the components which it carries. The interior of the reservoir is usually painted with a sealer that is resistant to the hydraulic fluid. These paints should be able to withstand the new synthetic hydraulic fluids, or the paint becomes loose and falls off.

The reservoir should be equipped with cleanout holes of ample size, so that the reservoir can be thoroughly cleaned without removing the reservoir cover. The reservoir should have an oil level gauge, so that the oil level can be visible. A return connection should be placed near the bottom of the reservoir and opposite the intake. Reservoirs may also be equipped with magnetized plugs or elements whose purpose is to pick up any minute particles of ferrous metals (iron or steel) that separate out of the system. Some reservoirs have an open top and others have a closed top. Those with an open top require a reservoir cover. These covers serve several purposes, such as:

1. Act as a closure to seal out the dirt.
2. Act as a base for the electric motor and pump. Sometimes these two components are carried on a subbase mounted on the reservoir cover.

131

Fig. 21. Cutaway view of a take-apart sump-type filter, showing the position of the magnetic rods.

Courtesy Marvel Engineering Company

3. Provide an inlet for the breather and the oil filler assembly. Also provide a hole for the return line from the relief valve.

The reservoir cover is usually made of heavy plate steel and painted to resist rust caused by moisture that collects on the bottom side of the cover.

THE INTAKE OR SUMP-TYPE FILTER

Most pumps are equipped with intake or sump-type filters. The symbol for the intake filter is F; another name for the intake filter is "strainer." The purpose of the intake filter is to keep dirt and other foreign matter from getting to the precision parts of the pump and causing damage. A take-apart sump-type filter is shown in Fig. 21. A cutaway view of the expanded metal protective housing shows the position of the magnetic rods. A filter loaded with ferric particles that were deliberately introduced into a hydraulic

Fig. 22. Filter loaded with ferric particles removed from a hydraulic system, illustrating the effectiveness of the magnetic field which completely surrounds the filter element.

Courtesy Marvel Engineering Company

system to illustrate the effectiveness of the magnetic field that completely surrounds the filter is shown in Fig. 22. Fig. 23 shows a sump-type filter that is mounted outside the reservoir, and is equipped with a warning signal that lets the operator know the condition of the filter. The intake filter is connected to the threaded end of the intake line, as shown in Fig. 24.

The filter is mounted in a horizontal position (see Fig. 24). This allows the filter to be mounted lower in the tank, so that no part of the filter is exposed to the air. If the screen section of the filter is exposed to air, the air enters the intake line; then it enters the pump, causing cavitation, which causes difficulties in the working parts of the pump and other important components throughout the hydraulic system.

Also note that the strainer should not touch the bottom of the reservoir. This prevents any dirt lying on the bottom of the reservoir becoming attached to the strainer and clogging it. This also causes cavitation. Lint particles in the oil often cause the filter to become clogged.

In some installations it is necessary to use filters with larger

133

Courtesy Marvel Engineering Company

Fig. 23. Sump-type filter equipped with a visual warning signal that indicates the condition of the filter to the operator.

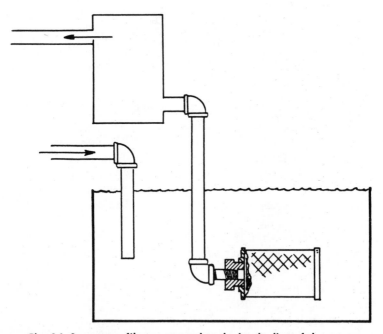

Fig. 24. Sump-type filter connected to the intake line of the pump.

mesh. If the reservoir is shallow, more than one filter may be necessary to provide the required filtering surface; keep the filters in the proper location with respect to the oil level and to the bottom of the reservoir.

Synthetic element filters are constructed to remove particles as small as one micron, which is 0.00004 inch in diameter. Syn-

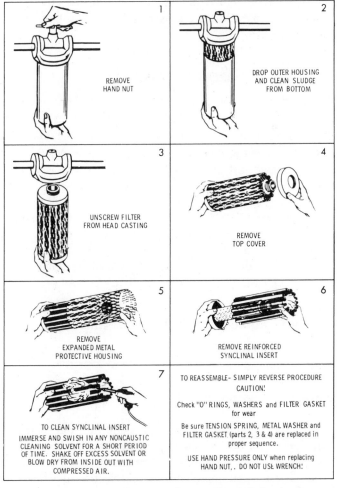

1. REMOVE HAND NUT

2. DROP OUTER HOUSING AND CLEAN SLUDGE FROM BOTTOM

3. UNSCREW FILTER FROM HEAD CASTING

4. REMOVE TOP COVER

5. REMOVE EXPANDED METAL PROTECTIVE HOUSING

6. REMOVE REINFORCED SYNCLINAL INSERT

7. TO CLEAN SYNCLINAL INSERT IMMERSE AND SWISH IN ANY NONCAUSTIC CLEANING SOLVENT FOR A SHORT PERIOD OF TIME. SHAKE OFF EXCESS SOLVENT OR BLOW DRY FROM INSIDE OUT WITH COMPRESSED AIR.

TO REASSEMBLE- SIMPLY REVERSE PROCEDURE CAUTION!

Check "O" RINGS, WASHERS and FILTER GASKET for wear

Be sure TENSION SPRING, METAL WASHER and FILTER GASKET (parts 2, 3 & 4) are replaced in proper sequence.

USE HAND PRESSURE ONLY when replacing HAND NUT, . DO NOT USE WRENCH!

Courtesy Marvel Engineering Company

Fig. 25. Maintenance of an oil filter.

thetic elements are made from a fiber that is bonded with a resinous material. The fiber-type filter elements are very porous, allowing large amounts of filter area for the size of the filter.

Most filters are easily cleaned by soaking them in solvent, then backwashing them with either air or solvent. The filter elements are usually easily replaced (Fig. 25).

Line-type filters are used in conjunction with the power device, but they are not used in the oil reservoir. A typical line-type filter is shown in Fig. 26.

BREATHER ASSEMBLY

The breather assembly is necessary for atmospheric pressure to reach the surface of the oil. If the top of the reservoir were sealed pressure tight, the pump could not function properly. The breather should allow the air to pass freely into the unit, but it must keep out all dirt and foreign matter. If a unit is located in an exceptionally dirty atmosphere, it may be necessary to use an automotive-type air cleaner to do a suitable job.

If, for any reason, it is necessary to remove the breather assembly, the hole should be plugged so that dirt cannot get into the unit; the pump should not be run, until the breather is replaced. Most breathers can be cleaned by soaking them in a solvent, then backwashing with compressed air.

DIAGNOSING PUMP FAILURE

A systematic check should be made to determine why a pump is not performing satisfactorily.

If the pump becomes noisy:

1. Check the alignment of the pump shaft with the shaft of the driving means. If the pump bearings, motor bearings, and couplings are getting hot, misalignment is indicated. To eliminate noise from misalignment, check the alignment very carefully.
2. Check for air leaks in the pump intake line. Take an oilcan

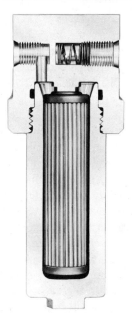

Fig. 26. Cutaway view of a high-pressure line type of filter.

Courtesy Marvel Engineering Company

and squirt oil on each joint while the pump is running. If this stops the noise, tighten the intake pipe connections.

3. Check for a clogged intake filter or for a filter that is too small. Noise from this is caused by cavitation in the pump. Removing the filter and running the pump for a short time without a filter can readily determine whether filter trouble is causing the noise. Do not run the pump without a filter if the oil is dirty or if there is sediment in the bottom of the reservoir. Then try a larger filter to determine whether the noise can be eliminated.

4. Check to determine whether the air breather assembly is clogged. Remove the assembly momentarily and determine whether pump noise ceases.

5. Check the pump speed to determine whether it is within the recommended range. If it is not, slow the pump speed to within the recommended range.

6. Check for foaming action of the oil, which causes air bubbles

137

to enter the system. Foaming may be caused either by the oil return lines being placed above the oil level or by the use of improper oil. Foaming is a condition that must be overcome. Addition of baffles may solve this problem.

7. Check the oil viscosity. If the oil is too heavy, it will not pump properly; thus causing considerable noise. The correct viscosity of oil is recommended by the pump manufacturer.

8. Check all pump gaskets and seals for air leaks. This can be done with an oilcan, squirting a little oil at each joint while the pump is running. Replace faulty gaskets and seals.

9. Check the relief valve for chatter. The pump may seem noisy when the noise is actually coming from the relief valve. A noisy relief valve often can be discovered by placing a hand on it to feel the vibration.

10. Check the pump for either broken or stuck parts. It is usually wise to check this last, as it is a major job to remove the pump from the unit and disassemble it.

11. Check the oil temperature and the pump temperature. When the pump becomes extremely hot, it may become very noisy as the working parts expand.

12. Check the size of the oil inlet line. It should be as large as the pump inlet port.

If the pump does not deliver the proper amount of oil or delivers no oil at all:

1. Check the direction of rotation of the pump shaft. Usually, an arrow indicating rotation direction can be found someplace on the pump. If the shaft is rotating in the wrong direction, this should be changed immediately. Shafts of some pumps may be rotated in either direction.

2. Check the oil inlet line for leaks. Leaks of sufficient size greatly hamper the pumping action.

3. Check the oil intake filter for lint, paint, and dirt. The filter should be cleaned if the proper flow of oil is to be expected from the pump.

4. Check for restrictions in the oil inlet line or for foreign substances in the line which accidentally may have entered the line during installation.

5. Check the oil level in the tank. Make certain that the intake line is well covered with oil and that the oil level is at the full gauge mark.
6. Check for a broken pump shaft or other broken pump parts which may cause the pump to fail to deliver oil.
7. Check the speed of the pump shaft. If it rotates too slowly, it may not deliver oil in the desired quantities.
8. If the oil is too heavy (viscosity too high), the pump may not pick up the prime. Select an oil that is recommended by the manufacturer of the pump.

If the pump does not develop the desired operating pressure:

1. Check for internal leakage in the pump caused by a torn pumping mechanism or by porosity within the pump body itself.
2. Check the setting of the relief valve. Turn down the pressure setting adjustment screw. Also check for a broken spring, stuck parts, excessive internal wear causing internal leakage, or porosity in the relief valve body.
3. Check for internal leakage in the system caused by a worn piston on the body in a valve; a worn piston or scored cylinder tube in a cylinder may cause so much leakage that the pump cannot pump enough liquid to overcome the leakage and build up pressure.
4. Check for open center valves in the system. Open center valves direct the flow of fluid to the reservoir when the valve spool is in the neutral position. Any normally closed in-line valve which is opened accidentally may be a source of trouble in the system.
5. Check for porosity or internal breakdown in the cored passages of the pump body. If any sign of porosity or broken passages is visible, change the pump body. Often, porosity is not visible, which presents a real problem. If possible, block off certain passages in the body and check them with compressed air.
6. Check to determine whether the pumping mechanism is performing satisfactorily. If the parts of the pumping mechanism stick, they do not perform properly and pressure cannot be developed. Dirt may be causing the mechanism to stick.

7. Check to determine whether the pump is delivering fluid. If it cannot deliver fluid, it cannot deliver pressure for the hydraulic system.

If the pump stalls:

1. Check to determine whether the electric motor or other driving means is large enough to take care of the pump at the pressure for which the relief valve is set. The relief valve may be set too high.
2. Check to see if the relief valve piston is stuck. This condition causes a pressure buildup until the pressure stalls the motor. This could break the pump housing or twist off the pump shaft.
3. Check for internal trouble in the pumping mechanism. Broken parts can jam the shaft, so that it cannot turn.

If there is external leakage around the pump shaft:

1. Check the alignment of the pump shaft. Excessive side thrust in coupling or pulley drives causes shaft packing trouble. Replace the packings immediately.
2. Check for excessive heat. This causes shaft seals to become brittle and fail.
3. Check the hydraulic fluid. Some fire-resistant types of hydraulic fluids cause packing failures.

INSTALLATION OF THE POWER DEVICE

There are several considerations in the installation of a hydraulic power device. Some of these are:

1. Install in a place that is accessible. Place it so that it is easily serviced. Keep the cleanout holes in the reservoir away from the wall or other obstruction.

   Since it may be necessary to change the setting on the relief valve occasionally, the unit should be set so that the operator can get to the relief valve. The pressure gauge should be located so that the operator can see it while he is working at the machine. This is not always possible, but it is desirable.

2. Installation of the power device in a pit is not advisable. First, it is difficult to service or to remove, and secondly, it does not allow ventilation which causes a build-up of heat and often causes power-unit failure. The reservoir has no chance to dissipate the heat.

3. If possible, install the power unit in the lowest point in the system. This allows the exhaust lines to drain naturally, and does not cause back pressure to build up in the system.

4. Mount the power device on a good solid foundation. Power devices weigh considerably, depending on their size, so make provisions for a good solid foundation.

5. Install the power device away from hot spots—the power device gets hot enough from its own internal heat without adding a high ambient temperature.

6. Install the power device so that the base is level. Install the power lines to the electric motor, if this is the means of driving the pump. Check with a level.

Before placing oil in a power device:

1. Be sure that the reservoir is clean. Remove one of the clean-out plates and shine a light into all of the compartments to make sure that no dirt, paper, or rags are in the reservoir.

2. Check all cleanout plate gaskets, pipe plugs, and pipe connections on the reservoir for tightness.

3. If synthetic fluids are to be used, check to see whether the inside of the reservoir is painted. If so, contact the manufacturer to see whether the paint is satisfactory for use with the synthetic fluid. Also check on the intake filter to see whether it is suitable for the fluid.

4. Use a hydraulic oil or fluid that is recommended by the pump manufacturer (see Chapter 8—Hydraulic Fluids).

Fill the reservoir to the full mark on the oil gauge. Be sure to replace the oil filler cap. This keeps dirt from getting into the

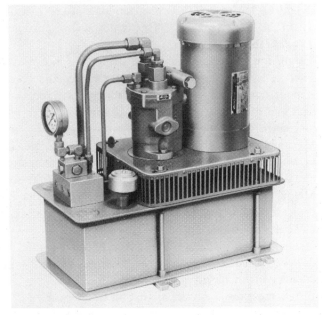

Courtesy Sperry Vickers, Division of Sperry Rand Corporation

Fig. 27. Vertical-mounted power package using a piston pump.

reservoir from the outside. Turn the relief valve adjustment screw outward, until the tension is released on the relief valve spring. Start the electric motor and check the direction of rotation of the pump shaft. If it is rotating in the wrong direction, stop the motor immediately. If the motor is a three-phase type, switch two leads to reverse the rotation of the motor shaft to the direction that it should rotate.

After making sure that the motor and the pump shafts are rotating in the correct direction, start turning downward on the relief valve adjusting screw. Turn it down slowly, and check the pressure gauge to determine whether the pressure reading increases. Turn down the relief valve screw until the desired operating pressure is reached. Check all joints and connections for pipe leaks.

Courtesy Double A, Division of Browne & Sharpe Mfg. Co.

Fig. 28. A power device with modular circuitry puts complete hyd-raulic circuitry on the power unit.

Power Devices With Valve Mounting Plates

In many applications it is desirable to mount the valves on the hydraulic power device, instead of locating them in various places over the machine. This provides a more compact installation, and, in most instances, makes it easier for the serviceman to make adjustments and repairs. It also eliminates some piping problems. Note the clean installation of power devices, as shown in Fig. 27 and Fig. 28.

To become better acquainted with the hydraulic power-unit circuit using *ANS* Graphic Symbols, note the circuit (Fig. 29A) and the photo (Fig. 29B). The circuit uses two variable-volume pumps.

143

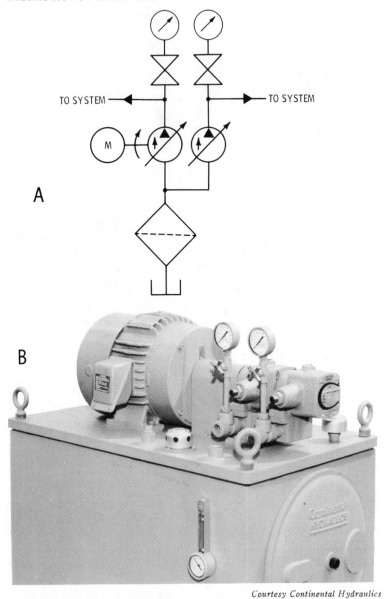

TO SYSTEM

TO SYSTEM

A

B

Courtesy Continental Hydraulics

Fig. 29. Hydraulic power-unit circuit and hydraulic power device: (A) Circuit with ANS Graphic Symbols; and (B) A hydraulic power device.

Courtesy Diamond Chain Company

Fig. 30. Installation of motor and pump coupling.

Courtesy Diamond Chain Company

Fig. 31. Installing the motor and pump coupling.

Courtesy Young Radiator Company

Fig. 32. Two-pass type of heat exchanger.

Installation of a Hydraulic Pump

Although the pump is already installed on the packaged power device, pointers on the installation of the pump are needed to bring out the steps required when replacements are necessary. Also, if a pumping device is to be built in the maintenance department, certain steps should be taken in the installation of the pump:

1. Place one-half of the motor and pump coupling on the pump shaft.

2. Mount the pump on a good flat solid base. Make certain that the pump shaft is in line with the shaft of the driving means. Check the coupling carefully after the pump mounting screws, containing lock washers, have been tightened securely (see Fig. 30, showing how the motor and pump coupling should be installed). Fig. 31 shows the parts of a coupling in an exploded view.

If the pump shaft and driving shaft are not properly aligned, considerable damage may be caused: the bearings may overheat; the seals may become worn; and considerable noise may be created. The couplings may become damaged quickly.

Fig. 33. A plunger type of pressure gauge.

Courtesy Scovill Fluid Power Division, Scovill Mfg. Co.

3. Hook up the intake line to the pump (see Fig. 24). Make sure that the intake line is large enough. The pipe should correspond to the size of the pipe thread on the intake port.
4. Hook up the pressure line. This line should be the same size as the pressure port on the pump.
5. Before starting the pump be sure that it is lubricated. Most pumps are equipped with grease fittings. Check the pump manufacturer's chart for type of lubricant to use. This lubricant is used to grease the pump bearings.

Heat Exchangers

This component is important where the ambient temperature is high, or where some condition within the hydraulic system creates considerable heat. Heat exchangers are not required on many hydraulic systems. If the temperature of the oil exceeds 150°F., it is advisable to use heat exchangers.

147

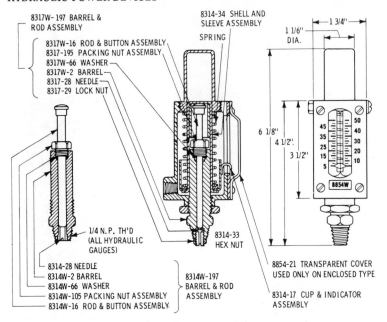

8317W-197 BARREL & ROD ASSEMBLY

8317W-16 ROD & BUTTON ASSEMBLY
8317-195 PACKING NUT ASSEMBLY
8317W-66 WASHER
8317W-2 BARREL
8317-28 NEEDLE
8317-29 LOCK NUT

8314-34 SHELL AND SLEEVE ASSEMBLY
SPRING

1 1/6" DIA.
1 3/4"

6 1/8"
4 1/2"
3 1/2"

45 50
35 40
25 30
15 20
5 10

8854W

1/4 N.P. TH'D (ALL HYDRAULIC GAUGES)

8314-33 HEX NUT

8314-28 NEEDLE
8314-2 BARREL
8314-66 WASHER
8314W-105 PACKING NUT ASSEMBLY
8314W-16 ROD & BUTTON ASSEMBLY

8314W-197 BARREL & ROD ASSEMBLY

8854-21 TRANSPARENT COVER USED ONLY ON ENCLOSED TYPE

8314-17 CUP & INDICATOR ASSEMBLY

8314W-197 USED IN 8314W, 8315W, 8316W, 8854W, 8856W, 9114W, 9115W HYDRAULIC GAUGES
8317W-197-USED IN 8317W, 8318W, 8319W, 8320W, 8857W, 8858W, 8859W, 8860W, 8862W, AND 9111W HYDRAULIC GAUGES

Courtesy Scovill Fluid Power Division, Scovill Mfg. Co.

Fig. 34. Cross-sectional diagram of a plunger-type pressure gauge.

When heat is caused either by ambient temperature or by a hot spot in the system, the exchanger should be connected to the exhaust line of the system. If the heat is caused by the oil spilling through the relief valve at high pressure, the exchanger should be connected to the exhaust of the relief valve.

Water is used as the cooling means in the heat exchanger. The flow of water should be controlled by a thermostat, so that when the oil cools to a certain temperature the water flow stops.

Heat exchangers may be either the single- or the double-pass type. In the single-pass type, the water should be piped in a direction opposite the flow direction of the oil, for the greatest heat transfer. On the double-pass type, the oil should enter the exchanger at the same end as the water enters and leaves. The basic parts of a two-pass type of heater exchanger may be studied in Fig. 32.

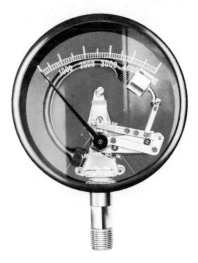

Courtesy Marshalltown Mfg., Inc.

Fig. 35. A common dial type of pressure gauge.

Other methods are used for cooling oil, such as a radiator-type exchanger that may use a fan for cooling—similar to the radiator and fan on an automobile.

Another method of cooling is to place a coil of copper tubing in the reservoir. Water is run through the tubing to effect a cooling action.

Pressure Gauges

The pressure gauge is an important item on the hydraulic power device. It registers the pressure set by the relief valve. After the pressure is set, the gauge line should be shut off to protect the gauge from pulsations (see Figs. 33, 34, and 35).

REVIEW QUESTIONS

1. What is the purpose of a hydraulic power device?

2. What are some of the advantages of gear-type pumps?

3. Describe a "hi-low" pump.

4. What function does the reservoir serve?

5. Why are baffles installed in a reservoir?

6. What is the function of an intake filter?

7. Why is a breather assembly used on a reservoir?

8. What causes a pump to become noisy?

9. What causes an abnormally low rate of flow from a pump?

10. What causes an abnormally low pump pressure?

11. What causes a pump to stall?

12. Where should a hydraulic power device be mounted?

13. What is the significance of a heat exchanger in a hydraulic system?

14. When are heaters used in a hydraulic system?

CHAPTER 8

Hydraulic Fluids

In hydraulics, hydraulic fluids are usually divided into three categories—petroleum-base fluids, synthetic-base fluids, and water. The first two fluids are used in "packaged-power devices." Water is generally used as the hydraulic fluid in central hydraulic systems.

The function of a good hydraulic fluid is threefold: (1) it is a means of transmission of fluid power; (2) it is a means of lubrication of the components of the fluid power system; and (3) it acts as a sealant. The selection of the proper hydraulic fluid is important, as it has a direct bearing on the efficiency of the hydraulic system, on the cost of maintenance, and on the service life of the system's components.

PETROLEUM-BASE FLUIDS

Three basic types of mineral oils are used: (1) Pennsylvania, or paraffin-base oils; (2) Gulf Coast, or naphthenic- and asphaltic-base oils; and (3) Mid-Continent, or mixed-base oils. These contain both naphthenic and paraffin compounds.

To obtain certain characteristics, chemicals are added to an oil. These chemicals are called *additives*. Additives cannot make an inferior oil perform as well as a good oil, but they can make a good oil perform even better. An additive may be in the form of an antifoam agent, a rust inhibitor, a film-strengthening agent, or an oxidation stabilizer.

The user should not attempt to place additives in a hydraulic oil. That job is primarily for the oil manufacturer or refiner.

SYNTHETIC-BASE FLUIDS

Since fire hazards are prevalent around certain types of hydraulically operated machines, especially where open fires are present, much research has been done to develop fire-resistant hydraulic fluids. These fluids are divided into two classifications —synthetic-base mixtures and water-base fluids. Not all synthetic-base fluids are fire resistant.

Synthetic-base fluids include chemical compounds, such as the chlorinated biphenyls, phosphate esters, or mixtures containing each. These hydraulic fluids are fire resistant, because a large percentage of phosphorous and chloride materials are included.

Water-base fluids depend on a high percentage of water to effect the fire-resistant nature of the fluid. In addition to water, these compounds contain antifreeze materials, such as glycol-type thickeners, inhibitors, and additives.

Synthetic-base fluids have both advantages and disadvantages. Some of the advantages are: (1) they are fire resistant; (2) sludge or petroleum gum formation is reduced; and (3) temperature has little effect on the thickening or thinning of the fluid. A disadvantage of many synthetic fluids is their deteriorating effect on some materials, such as packings, paints, and some metals used in intake filters.

QUALITY REQUIREMENTS

Certain qualifications are demanded in a good hydraulic oil—an oil should not break down and it should give satisfactory service. Some of these requirements are:

1. Prevent rusting of the internal parts of valves, pump, and cylinders.
2. Prevent formation of a sludge or gum which can clog small passages in the valves and screens in filters.
3. Reduce foaming action which may cause cavitation in the pump.
4. Properties that provide a long service life.
5. Retain its original properties through hard usage—must not deteriorate chemically.

6. Qualities which resist changing the flow ability or viscosity as the temperature changes.

7. Form a protective film which resists wear of working parts.

8. Prevent pitting action on the parts of pumps, valves, and cylinders.

9. Does not emulsify with the water that is often present in the system either from external sources or from condensation.

10. No deteriorating effect on gaskets and packings.

MAINTENANCE

Proper maintenance of a hydraulic oil is often forgotten. Too often, hydraulic oil is treated as matter-of-fact. A few simple rules regarding maintenance are:

Fig. 1. Store hydraulic oil in a clean container.

1. Store oil in a clean container (Fig. 1). The container should not contain lint or dirt.

2. Keep lids or covers tight on the oil containers (Fig. 2), so that dirt or dust cannot settle on the surface of the oil. Oil should never be stored in open containers.

3. Store oil in a dry place; do not allow it to be exposed to rain or snow (Fig. 3).

4. Do not mix different types of hydraulic oils. Oils having different properties may cause trouble when mixed (Fig. 4).

5. Use a recommended hydraulic fluid for the pump (Fig. 5).

6. Use clean containers for transporting oil from the storage tank to the reservoir.

7. Make sure that the system is clean before changing oil in the power unit; do not add clean oil to dirty oil.

8. Check the oil in the power unit regularly. Have the oil supplier check a sample of the oil from the power unit in his laboratory. Contaminants often cause trouble—these can be detected by frequent tests, which may aid in determining their source. On machines that use coolants or cutting oil,

Fig. 2. Keep covers tight on oil containers.

extreme caution should be exercised to keep these fluids from entering the hydraulic system and contaminating the oil.

Fig. 3. Store hydraulic oil in a dry place.

9. Drain the oil in the system at regular intervals. It is difficult to set a hard and fast rule as to the length of the interval. In some instances, it may be necessary to drain the oil only every two years; however, once each month may be necessary for other operating conditions. This depends on operating conditions and on the original quality of the hydraulic oil. Thus, several factors should be considered in determining the length of the interval.

Before placing new oil in the hydraulic system, it is often recommended that the system be cleaned with a hydraulic system cleaner. The cleaner is placed in the system after the oil has been

155

Fig. 4. Do not mix different types of hydraulic oils.

Fig. 5. Use a recommended hydraulic fluid for the
hydraulic pump.

removed. The hydraulic system cleaner should be used while the hydraulic system is in operation, and usually requires 50 to 100 hours to clean the system. Then, the cleaner should be drained; the filters, strainers, and oil reservoir cleaned; and the system filled with a good hydraulic oil.

If hydraulic oil is spilled on the floor in either changing or adding oil to the system, it should be cleaned up at once. Good housekeeping procedure is important in reducing fire and other safety hazards.

CHANGE OF FLUIDS IN A HYDRAULIC SYSTEM

If the fluid in a hydraulic system is to be changed from a petroleum-base fluid to a fire-resistant fluid—or vice versa, the system should be drained and cleaned completely.

In changing from a petroleum-base fluid to a water-base fluid:

1. Drain out all the oil—or at least as much as possible, and clean the system.

2. Either remove lines which form pockets, or force the oil out with a blast of clean, dry air.

3. Strainers should be cleaned thoroughly. Filters should be cleaned thoroughly and the filter element replaced.

4. Check the internal paint in all components; it is likely that the paint should be removed.

5. Check the gaskets and packings; those that contain either cork or asbestos may cause trouble.

6. Flush out the system. Either a water-base fluid or a good flushing solution is recommended. Carbon tetrachloride is not recommended, because it tends to form hydrochloric acid by reacting with the water. This results in corrosion.

7. Since hydraulic fluids are expensive, the system should be free of external leaks.

In changing from a water-base fluid to a petroleum-base fluid:

1. Remove all of the water-base fluid. This step is very important. A small quantity of water-base fluid left in the system can cause considerable trouble with the new petroleum-base fluid.

2. The reservoir should be scrubbed and cleaned thoroughly. If the interior of the reservoir is not painted, it should be coated with a good sealer that is not affected by hydraulic oil.

3. The components should be dismantled and cleaned thoroughly. Cleaning with steam is effective.

4. Flush the system with hydraulic oil and then drain.

5. Fill the system with a good hydraulic oil.

A similar procedure should be used in changing from synthetic-base fluids to petroleum-base fluids—and vice versa. If phosphate-base fluids are used, the packings should be changed. If satisfactory performance from a hydraulic fluid and a hydraulic system is expected, use a good grade of hydraulic fluid, keep it clean, change at regular intervals, do not allow it to become overheated, and keep contaminants out of the system.

SELECTION OF A HYDRAULIC FLUID

The main functions of the hydraulic fluid are to transmit a force applied at one point in the fluid system to some other point in the system and to reproduce quickly any variation in the applied force. Thus, the fluid should flow readily, and it should be relatively incompressible. The choice of the most satisfactory hydraulic fluid for an industrial application involves two distinct considerations: (1) the fluid for each system should have certain essential physical properties and characteristics of flow and performance; and (2) the fluid should have desirable performance characteristics over a period of time. An oil may be suitable when initially installed; however, its characteristics or properties may change, resulting in an adverse effect on the performance of the hydraulic system.

The hydraulic fluid should provide a suitable seal or film between moving parts, in order to reduce friction. It is desirable that the fluid should not produce adverse physical or chemical changes while in the hydraulic system. The fluid should not promote rusting or corrosion in the system, and it should act as a suitable lubricant to provide film strength for separating the moving parts to minimize wear between them.

Certain terms are required to evaluate the performance and suitability of a hydraulic fluid. Important terms are discussed in the paragraphs that follow.

Specific Weight

The term *specific weight* of a liquid indicates the weight per unit of volume. For example, water at 60°F. weighs 62.4 pounds per cubic foot. The "specific gravity" of a given liquid is defined as *the ratio of the specific weight of the given liquid divided by the specific weight of water*. For example, if the specific gravity of an oil is 0.93, the specific weight of the oil is (0.93×62.4), or approximately 58 pounds per cubic foot. For commercially available hydraulic fluids, the specific gravity may range from 0.80 to 1.45.

Viscosity

Viscosity is a frequently used term. In many instances, the term is used in a general, vague, and loose sense. To be definite and specific, the term "viscosity" should be used with a qualifying term.

The term *absolute or dynamic viscosity* is a definite specific term. As indicated in Fig. 6, the hydraulic fluid between two parallel plates adheres to the surface of each plate, which permits

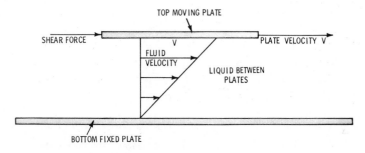

Fig. 6. Diagram illustrating the shearing action of a liquid.

one plate to slide with respect to the other plate (as playing cards in a deck); this results in a "shearing" action in which the fluid layers slide with respect to each other. A "shear" force acts to "shear" the fluid layers at a certain velocity, or rate of relative motion, to provide the shearing action between the layers of fluid. The term "absolute or dynamic viscosity" is a physical property of the hydraulic fluid, which indicates the ratio of the shear force and the rate or velocity at which the fluid is being sheared.

To simplify, a very *viscous* fluid or a fluid having a high dynamic viscosity is a fluid that does not flow freely, or fluid hav-

159

ing a low dynamic viscosity flows freely. The term *fluidity* is the reciprocal of "dynamic viscosity." A fluid having a high dynamic viscosity has a low fluidity, and a fluid having a low dynamic viscosity has a high fluidity. In general, the dynamic viscosity of a liquid decreases as temperature increases; therefore, as an oil is heated, it flows more freely. Because of pressure effects, it is difficult to draw general, firm conclusions for all oils. It is possible for an increase in fluid pressure to increase the viscosity of an oil.

Saybolt Universal Viscometer

The term "dynamic viscosity" is sometimes confused with the reading taken from the Saybolt Universal Viscosimeter. In actual industrial practice, this instrument has been standardized arbitrarily for testing of petroleum products. Despite the fact that it is called, a viscosimeter, the Saybolt instrument does not measure "dynamic viscosity." A diagram illustrating the Saybolt viscosimeter is shown in Fig. 7.

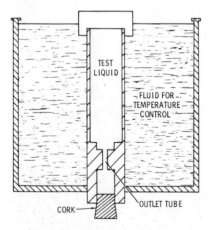

Fig. 7. Diagram illustrating the basic operating principle of the Saybolt viscosimeter.

In operating the instrument, the liquid to be tested is placed in the central cylinder, which is a short, small-bore tube having a cork at its lower end. Surrounding the central cylinder, a liquid bath is used to maintain the temperature of the liquid that is being tested. After the test temperature has been reached, the cork is pulled, and the time in seconds, that is required for 60 milliliters of the test fluid to flow out of the cylinder is measured with a stop watch. This measured time, in seconds, is called the *Saybolt Universal Reading.*

The *S.A.E.* (Society of Automotive Engineers) has established standardized numbers for labeling of the oils. For oils tested at 130°F. in a standard Saybolt Universal instrument, Table 1 indicates *S.A.E.* numbers for the corresponding ranges of Saybolt Universal readings.

Table 1. Range Of Saybolt Readings, Seconds

SAE Numbers	Minimum	Maximum
10	90	less than 120
20	120	less than 185
30	185	less than 255

For example, if an oil is labeled "SAE 10," the Saybolt Universal reading at 130°F. is in the range, from 90 to less than 120 seconds.

Viscosity Problems

If the viscosity of the hydraulic fluid is *too high* (fluid does not flow as freely as desired), the following undesirable actions may result:

1. Internal resistance, or fluid friction, is high, which means a high resistance to flow through the valves and pumps.

2. Power consumption is high, because fluid friction is high.

3. Fluid temperature is high, because friction is high.

4. Pressure drop through the system may be higher than desired, which means that less useful pressure is available for doing useful work.

5. The motion and operation of the various parts may be slow and sluggish as a result of the high fluid resistance.

If the viscosity of the hydraulic fluid is *too low* (fluid flows more freely than desired), the following undesirable actions may result:

1. More leakage may occur in the clearance space than is desired.

2. A lower pressure may occur in the system than is desired.

3. An increase in wear may occur, because of the lack of a strong fluid film between mechanical parts that move in relation to each other.

4. Pump leakage may increase, resulting in reduced pump delivery and efficiency.

5. A loss of control may occur, because fluid film strength is reduced.

With respect to Saybolt readings, the viscosimeter readings of oils in service should not exceed 4000 seconds, and they should not read less than 45 seconds.

Viscosity Index

Ideally, the dynamic viscosity of an oil should change only slightly, as the temperature changes. In the automobile engine, the oil in the crankcase is operated over a wide range of temperatures. On a very cold winter morning, after the car has been operated for some length of time, the temperature of the oil may be very low, and the dynamic viscosity of the oil may be very high. If the dynamic viscosity of the oil is excessively high, large forces and large amounts of power may be required to "shear" the oil films. Also, after the engine has been operated for a period of time on a hot summer day, the temperature of the oil may be very high, and the dynamic viscosity of the oil may be too low; therefore, the oil may not form a suitable lubricating film between the sliding surfaces. A breakdown of the oil film may result in excessive wear of the metal surfaces and a loss of power in the engine.

The term *viscosity index* is an arbitrarily defined ratio; it indicates the relative change in Saybolt Universal reading, with respect to temperature. The most desirable oils are those that have a high viscosity index; that is, the change in Saybolt reading is relatively small, as the temperature changes. Oils having a small viscosity index register a relatively large change in Saybolt reading, as the temperature changes.

Lubricating Value

The terms *oiliness* and *lubricity* are used to refer to the lubricating value of an oil. These terms are most often used when the moving surfaces are relatively close and may make metal-to-metal contact. At the same pressure and temperature, an oil *A* may be a better lubricant than another oil *B*; therefore, oil *A* possesses more "oiliness" or "lubricity" than oil *B*. The lubricating value of a fluid depends on its chemical structure and its reaction with various metal surfaces when the metal surfaces are relatively close to each other. Thus, oiliness and lubricity are extremely important in the performance of an oil.

Pour Point

The *pour point* of a fluid is defined as the lowest temperature at which the fluid flows when it is chilled under given conditions. The pour point is important when the hydraulic system is exposed to low temperatures. As a general rule, the most desirable pour point should be approximately 20°F. below the lowest temperature to which the fluid will be exposed.

Oxidation and Contamination

Oxidation is a chemical reaction in which oxygen combines with another element. As the air contains oxygen, the oxygen that is involved in fluid oxidation comes from exposing or mixing the fluid with air. The oxidation reaction increases with the increased exposure of the oil to air.

Undesirable quantities of air in hydraulic systems can be due to mechanical causes, such as air leakage into the oil suction line, low fluid level in the oil reservoir, and leakage around the packing. Air leakage may result in the erratic motion of mechanical parts, and it also may cause the fluid to oxidize more rapidly. All oils contain some air in solution, which may not cause any trouble. If the air is not in solution, a foaming action may result. If trapped in a cylinder, air that is not in solution is highly compressible; however, the oil is not as highly compressible as the air.

Irregular action of a cylinder, for example, may result if a significant quantity of air becomes undissolved.

Ferrous metals are destroyed by rust. Rust can develop in a hydraulic system if moisture is present; this moisture may be the result of condensation from air that enters through leaks on the intake (low pressure) side of a pump.

The "oxidation stability" of an oil refers to the inherent ability of an oil to resist oxidation. Oxidation increases with increases in temperature, pressure, and agitation. Oxidation also increases as the oil becomes contaminated with such substances as grease, dirt, moisture, paint, and joint compound. Various metals also promote oil oxidation, and the various fluids have different oxidation characteristics.

Table 2 lists the essential properties of the commercially available hydraulic fluids.

Table 2. Properties of Available Hydraulic Fluids

Petroleum-Base Fluids
Viscosity range, Saybolt Universal reading, in   seconds, at 100°F .... 40 to 5000
Operating temperature, in °F .... −75 to 500
Minimum viscosity index .... 76 to 225

Fire-Resistant Fluids (Water-oil emulsions, water-glycol, phosphate-ester, chlorinated hydrocarbon, silicate ester, silicon)
    Viscosity range, Saybolt Universal reading, in
        seconds, at 100°F .... 20 to 5000

Operating temperature, in °F .... −100 to 600

REVIEW QUESTIONS

1. List three disadvantages of using water in a fluid power system.

2. What may be the cause of hydraulic oil becoming overheated in a hydraulic system?

3. In what ways can air enter a hydraulic system?

4. In what ways can dirt get into a hydraulic system, despite the fact that a suitable filter is employed in the system?

5. What precautions should be taken in changing the oil in a hydraulic system?

6. In storing hydraulic fluids, what precautions should be exercised?

7. What hazards are presented when hydraulic oil remains on the floor?

8. List three types of commonly used hydraulic fluids.

9. What is the effect of some fire-resistant fluids on packing, gaskets, filters, etc.

CHAPTER 9

Piping

The function of the piping in either a hydraulic or a pneumatic system is to act as a leakproof carrier of the fluid. The piping in a fluid power system may be compared to the water piping in a home: one section provides water to the bathroom; a second section provides water to the kitchen sink; and still another section provides the water for an automatic washer, a dishwasher, or a lavatory.

The piping or "plumbing" in a fluid power system is too often an afterthought, with the result that it is often a source of trouble. It is important that the piping in any fluid power system should be properly arranged to provide maximum efficiency and trouble-free service.

CLASSIFICATION

Piping may be divided into three classes: (1) rigid; (2) semi-rigid, or tubing; and (3) flexible, or hose.

Rigid Pipe

Rigid steel pipe is available in four weights as follows:

1. Standard (STD), or Schedule 40.

2. Extra strong (XS), or Schedule 80, for 1000 *psi*.

3. Schedule 160, for 3000 *psi*.

4. Double extra strong (XXS).

Pipe sizes are specified by the *nominal inside diameter*—as ¼, ½, ¾, 1, and 1 ¼ inches. All weights of one size are of the same outside diameter, but the actual inside diameter varies, depending on the wall thickness.

Fittings used in connection with steel pipe are elbows, street elbows, crosses, tees, and unions. The sizes of the fittings correspond to the pipe sizes.

Semirigid (Tubing)

Steel, aluminum, and copper seamless tubing are all used for oil and air systems. These can be grouped as:

1. Seamless steel (S.A.E. 1010), fully annealed.

2. Stainless steel, seamless (18-8), fully annealed, suitable for bending and flaring.

3. Aluminum, seamless (B50S-0).

4. Copper, seamless, fully annealed.

Since tubing can be bent, lines from tubing may require a minimum number of fittings. Tubing is specified by outside diameter and wall thickness (see Table 1). Copper tubing is not usually recommended for hydraulic systems.

Fittings that are used in connection with tubing are: elbows, crosses, tees, unions, and connectors (Fig. 1). Tube fittings can be classified into two general groups: (1) flared fittings (Fig. 2); and (2) flareless fittings (Fig. 3). A stainless steel high-pressure fitting that can be used for pressures from 10,000 to 20,000 *psi* is shown in Fig. 4. The flared tubing should be flared properly. Two different flare angles are used for flared tubing. The *USASI* standards call for a 37° flare, and *S.A.E.* standards specify a 45° flare. Check the flare on the fitting before flaring the tubing.

Flexible Piping (Hose)

Flexible hose is available for many types and classes of work. Hose is usually specified by the inside and outside diameters (Fig. 5). The so-called "tube" is the lining or part that comes into actual contact with the fluid or material being handled. The "car-

Table 1. Safe Internal Working Pressure For Tubes

(Cold-Drawn Seamless Steel)
Soft Annealed
Yield Point 30,000 Ultimate 48,000
Rockwell—B-50

Wall Thickness	¼	⁵⁄₁₆	⅜	⁷⁄₁₆	½	⅝	¾	⅞	1
.028	2240	1795	1493	1281	1120	896	747	640	560
.035	2800	2244	1867	1602	1400	1120	933	800	700
.042	3360	2692	2240	1922	1680	1344	1120	960	840
.049	3920	3141	2613	2243	1960	1568	1307	1120	980
.058	4640	3718	3093	2654	2320	1856	1547	1326	1160
.065	5200	4167	3467	2975	2600	2080	1733	1486	1300
.072	5760	4615	3840	3295	2880	2304	1920	1646	1440
.083	6640	5321	4427	3799	3320	2656	2213	1897	1660
.095	7600	6090	5067	4348	3800	3040	2533	2171	1900
.109		6987	5813	4989	4360	3488	2907	2491	2180
.120		7692	6400	5492	4800	3840	3200	2743	2400
.134			7147	6133	5360	4288	3573	3063	2680

cass" is the supporting structure of the hose; it lies between the tube and the cover. The carcass material may be cotton, synthetic fiber, asbestos, or wire; and it may be woven, braided, wrapped or wound spirally. The "cover" is the outside covering element of the hose. The purpose of the cover is to protect the carcass from abrasion or other destructive forces, pulsating pressures, falling objects, sun rays, weather, oils, greases, acids, and chemicals. Several classes of pressure are used in classifying the types of hose. The "recommended working pressure" is that pressure at which a given hose may be operated safely for satisfactory service. The "test pressure" is the pressure that a hose is guaranteed to withstand. The "burst pressure" is the pressure at which the hose is ruined and rendered unfit for further service.

Hose should be oil resistant; this requirement is clear for all hydraulic systems. For air or pneumatic systems, it should be noted that oil vapor in the lines from the compressor and the oil are introduced into the lines that are connected to pneumatic tools. The carcass of the hose should be strong enough for the intended service, and the cover should be rugged enough to withstand hard abrasive wear. The hose should be flexible and easy to handle. Various end fittings are available for the different types of hose.

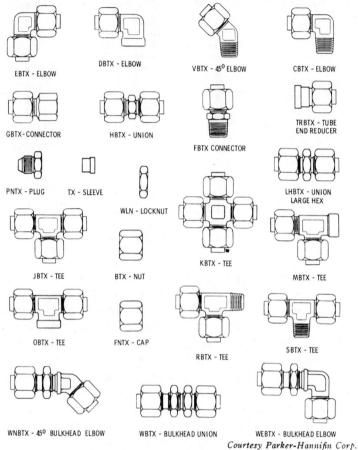

Courtesy Parker-Hannifin Corp.

Fig. 1. Tube fittings used in fluid power systems.

GENERAL FEATURES OF PIPING INSTALLATION

Several rules are important in installation of piping. Some of these rules are:

1. The piping should be clean. Be sure that there is no scale inside the pipe.

2. The piping should have a cross-sectional area that is large enough for the fluid to pass with low resistance. The di-

Fig. 2. A flared-type tube fitting.

Courtesy Parker-Hannifin Corp.

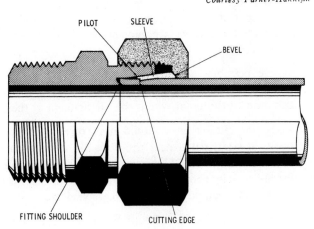

PILOT   SLEEVE

BEVEL

FITTING SHOULDER   CUTTING EDGE

Courtesy The Weatherhead Company

Fig. 3. A flareless tube fitting.

171

ameter of the pipe is usually governed by the port size of the components to which it is connected. In oil flow, flow velocity is usually kept below 15 feet per second.

3. The piping should be as short as possible with the least number of bends. Sharp bends should be eliminated whenever possible.

4. When pipe is installed, it should be fastened securely. Brackets and hangers should be used to prevent vibration. In some instances of excessive vibration, it is often helpful to insert a section of flexible hose (Fig. 6).

PLANNING A COMPRESSED-AIR DISTRIBUTION SYSTEM

A drop in pressure between the compressor and the point of use of the compressed air is a loss that cannot be recovered. The distribution system in a compressed-air plant is an important factor. The *Compressed Air Handbook* (McGraw-Hill, New York. 1954, Pages 3-6) states that the following general rules should be observed in planning the compressed-air distribution system:

1. Pipe sizes should be large enough that the pressure drop between the receiver and the point of use does not exceed 10 percent of the initial pressure. Provision should be made not only for present air capacity but also for reasonable future growth.

2. Where possible, use a loop system around the plant and within each shop and building. This gives a two-way distribution to the point where the air demand is the greatest.

3. Long distribution lines should have receivers that are adequate in size located near the far ends, or at points of occasional heavy use. Many peak demands for compressed air are instantaneous, or of short duration, and storage capacity near such points avoids excessive drop.

4. Each header or main should be provided with outlets as close as possible to the point of use. This permits shorter hose lengths and avoids large pressure drops through the

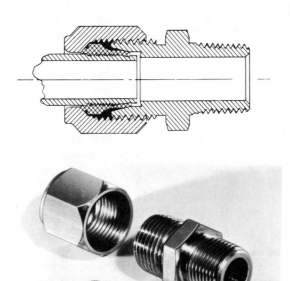

Courtesy Parker-Hannifin Corp.

Fig. 4. A stainless steel high-pressure fitting for pressures ranging from 10,000 to 20,000 psi.

Courtesy The Weatherhead Company

Fig. 5. Illustrating the various layers of material in a flexible hose.

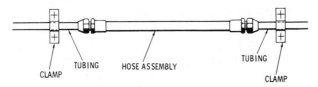

Fig. 6. Illustrating use of flexible hose to absorb excessive vibration.

173

hose. Outlets should be located at the top of the pipe line to prevent carry-over of condensed moisture to the tool.

5. Arrange all piping so that it slopes toward a drop leg or moisture trap in order that condensed moisture may be removed to prevent it being carried into the air tools or air-operated devices where it is harmful. The slope of the air lines should always be away from the compressor to prevent condensate from draining backward into the compressor cylinder.

INSTALLATION OF RIGID PIPE

When rigid pipe is used for an installation, standard piping should be used for pneumatic systems, and either Schedule 80 or Schedule 160 should be used for hydraulic systems. Schedule 80 is normally used for hydraulic systems requiring pressures up to 1000 pounds per square inch, and Schedule 160 is used for pressures up to 3000 pounds per square inch.

The ends of steel pipe may be threaded with a pipe die. Also, the ends of the pipe may be welded to a welding flange (Fig. 7), which fits a mating connection on a valve or cylinder. Care should be exercised in threading the pipe to be certain that the pipe threads are cut properly. After the pipe threading is finished, all pipe shavings, threading compound, and burrs should be removed from both inside and outside the pipe. Also, absolutely no rust or dirt should remain inside the pipe. When welding flanges are used, be sure that all loose welding beads are removed from the interior of the pipe.

INSTALLATION OF SEMIRIGID TUBING

After determining the required length of tubing, the next step is to cut the tubing. It is important to do the job properly; care should be taken not to distort, flatten, or nick the tubing while it is being cut. The final cut should be square, smooth, and free from external and internal burrs. A tube cutter is advantageous in making a proper cut (Fig. 8 and Fig. 9). Remove the burr from both the outside and the inside portions of the tube. If the tubing is cut with a hacksaw, the end of the tubing should be

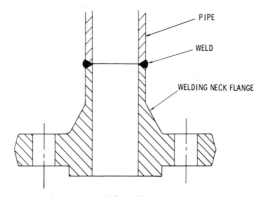

Fig. 7. A welding flange connection.

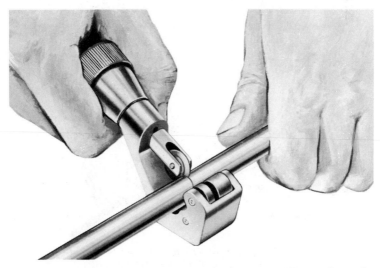

Courtesy Imperial-Eastman Corporation

Fig. 8. A typical tube cutter.

filed (Fig. 10). After preparing the end of the tube, the tube should be cleaned. Blowing compressed air through the tube is one method of cleaning it.

If flared fittings are used, it is, of course, necessary to flare the tubing. The flaring operation can be accomplished by means of a mechanical flaring device or machine; however, hand-type flaring

175

Fig. 9. Another type of tube cutter.

Courtesy Parker-Hannifin Corp.

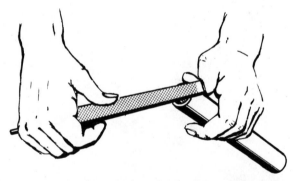

Fig. 10. If tubing is cut with a hacksaw, the end of
the tubing should be filed.

tools are commonly used. Some of these are the hammer-type flaring tool (Fig. 11), the vise block with flaring pin (Fig. 12), and the hand vise and clamp-type, or combination flaring tool (Fig. 13). Fig. 14 shows a flared tube that was not cleaned properly before the tool was applied. A split flare, which may be caused by incorrectly hardened tubing, by uneven texture of the tube, or by scratches or draw marks opening up, is shown in Fig. 15. The

Fig. 11. A hammer-type flaring tool for tubing.

manufacturer's instructions for using each tool should be followed closely. The gripping and positioning of the tubing in the clamp are very important.

The tubing can be bent either by hand operation or by machine. A crank-operated tube bender and a lever-type hand-operated tube bender are shown in Figs. 16 and 17. A hand-operated tube bender clamped in a vise and a gear-type tube bender are illustrated in Figs. 18 and 19. Bends should not be too sharp; the radius of the smallest bend (from center line to center line of the tubing) should measure approximately three times the outside diameter of the tubing. Care should be taken to avoid wrinkling and flattening the tubing (Fig. 20).

Tubing should be installed carefully. Careful planning beforehand may avoid real difficulties. Straight line connections, particularly short lengths, should be avoided wherever possible. Care should be taken to eliminate stress on tubing; long tubing should be supported by suitable brackets. The tubing should not be used

Fig. 12. Illustrating the vise block with flaring pin.

Courtesy Parker-Hannifin Corp.

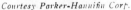

Fig. 13. The hand vise and clamp-type, or combination, flaring tool.

Courtesy Imperial-Eastman Corporation

to support other devices. A number of correct and incorrect installations are shown in Fig. 21. The face of the flare and the seat of the fitting should be clean and free from dirt when flared fittings are used.

Two types of pipe threads may be used on tube fittings, or piping: (1) the *National Taper Pipe Threads,* and (2) the *Dryseal*

178

Fig. 14. Result of improper cleaning before tubing was flared.

Fig. 15. A split flare.

Fig. 16. Crank-operated tube bender.

179

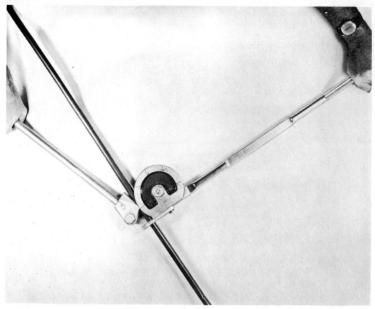

Courtesy Imperial-Eastman Corporation

Fig. 17. Lever-type hand-operated tube bender.

Taper Pipe Threads. These threads are completely interchangeable. *Dryseal* pipe thread joints provide less leakage than National Taper pipe threads. Each type of thread can be used to form a seal with the other type. It is necessary to use a sealing compound when one of the threads is not the *Dryseal* type of thread. Tube fittings that have straight threads and an "O" ring seal are often used.

It is good practice to lubricate pipe threads before assembly. Any light oil may serve as a lubricant. A sealer, rather than a lubricant, should be used on types of pipe threads other than *Dryseal*. White lead, *Bakelite*, varnish, *Petrolatum*, and a number of other sealers may be used.

Teflon tape is now being used as a thread seal on many applications. Several advantages are: easy to apply, neat in appearance, inexpensive, no damage to the connection, and can withstand temperatures up to 600°F. *Teflon* tape is similar to white adhesive tape in appearance.

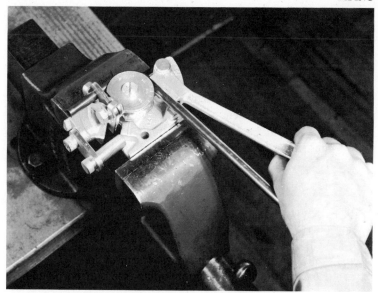

Courtesy Parker-Hannifin Corp.

Fig. 18. Hand-operated tube bender clamped in a vise.

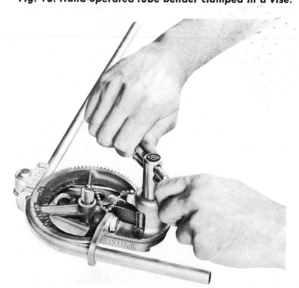

Courtesy Imperial-Eastman Corporation

Fig. 19. Gear-type tube bender.

Several precautions should be observed in installing tubing (see Fig. 21). Several of these are:

1. Avoid straight line connections wherever possible, especially in short runs.
2. Design the piping systems symmetrically. They are easier to install and present a neater appearance.
3. Care should be taken to eliminate stress from the tubing lines. Long tubing should be supported by brackets or clips.

Fig. 20. Avoid wrinkles and flattening in tube bending.

Courtesy Parker-Hannifin Corp.

All parts installed on tubing lines, such as heavy fittings, valves, etc., should be bolted down to eliminate tubing fatigue.

4. Before installing tubing, inspect the tube to make certain that it conforms to the required specifications, that it is of

the correct diameter, and that the wall thickness is not out of round.

5. Cut the ends of tubes reasonably square. Ream the inside of the tube and remove the burrs from the outside edge. Excessive chamfer on the outside edge destroys the bearing of the end of the tube on the seat of the fitting.

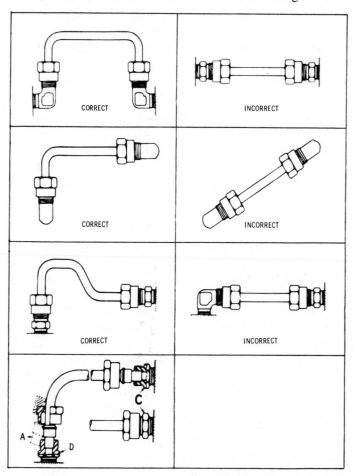

Courtesy The Weatherhead Company

Fig. 21. Correct method (left) and incorrect method (right) of installing tubing in a system.

183

6. To avoid difficulty in assembly and disconnecting, a sufficient straight length of tube must be allowed from the end of the tube to the start of the bend. Allow twice the length of the nut as a minimum.

7. Tubes should be formed to assemble with true alignment to the center line of the fittings, without distortion or tension. A tube which has to be sprung (see position *A* in Fig. 21) to be inserted into the fitting has not been properly fabricated, and when so installed and connected, places the tubing under stress.

8. When assembling the tubing, insert the longer leg to the fitting as at *C* (Fig. 21), and then insert the other end into fitting *D*. Do not screw the nut into the fitting at *C*. This holds the tubing tightly, and restricts any movement during the assembly operation. With the nut free, the short leg of the tubing can be moved easily, brought to position properly, and inserted into the seat in the fitting *D*. The nuts can then be tightened as required.

INSTALLATION OF FLEXIBLE PIPING

Correct and incorrect ways of installing hose are illustrated in Fig. 22. Hose assemblies, that is, the hose and the fittings on the hose are of two types: (1) those that are permanently assembled at the factory, and (2) those with reusable fittings. A method in which reusable fittings are used to make a hose assembly is illustrated in Fig. 23, and hose assemblies with reusable fittings are shown in Fig. 24.

The illustrations in Fig. 22 may be used as a guide for the proper methods of installing hose assemblies. The hose should not be bent or twisted too sharply, and it should not be placed in torsion at any time during installation. Sharp or excessive bends may cause the hose to kink or rupture.

In straight-hose installations (see Fig. 22), enough slack should be allowed in the hose line to provide for changes in length that occur when pressure is applied. This change in length can range from plus two percent to minus four percent.

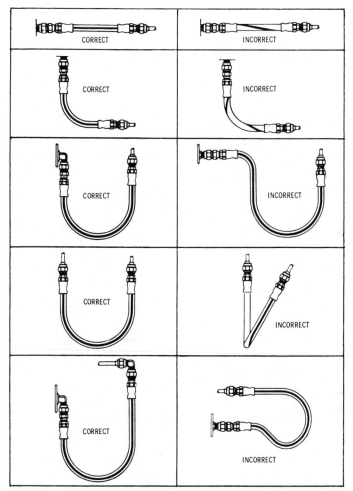

Courtesy The Weatherhead Company

Fig. 22. Correct way (left) and incorrect way (right) to install hose.

Design the installation so that the hose assembly is accessible for inspection or easy removal. It should be remembered that the metal end fittings cannot be considered as part of the flexible portion of the hose assembly. Most hoses shorten approximately five percent in length when pressure is applied. A "slack" allowance for change in length under working conditions should be made.

185

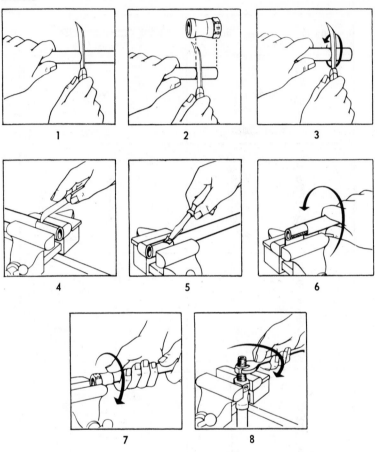

Courtesy The Weatherhead Company

Fig. 23. Illustrating a method of using reusable fittings in making up a hose assembly.

The installation should be studied to determine where pipe elbows can be used to advantage. Strain on the hose assembly can be relieved by using elbow tubing and pipe fittings.

The bend radius is important. A good working rule is that the bend radius should be five, or more, times the outside diameter of the hose. Various types of assemblies may be made up, using hose and tubing (Fig. 25). Such assemblies are often used on

Courtesy Parker-Hannifin Corp.

Fig. 24. Hose assemblies with reusable fittings.

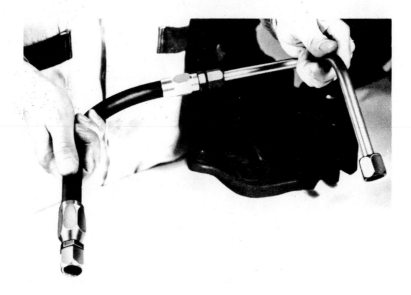

Courtesy Parker-Hannifin Corp.

Fig. 25. Tube and hose assembly.

equipment that is mass produced. The fitting and hose assembly shown in Fig. 26 has various applications on fluid power equipment. This assembly can be made quickly on the job site.

187

MANIFOLDS

The intricate systems that are now in use emphasize compactness; therefore, many manifolds are employed, especially on hydraulic systems. It is often desirable to mount all the valves on a manifold panel. This eliminates piping between the valves, and reduces the possibility of leaks. Although manifolds are relatively expensive, they greatly reduce assembly time in connecting up a system. Various designs are used in developing manifolds; these include the drilled type, the bolted-on type, and the brazed-plate type.

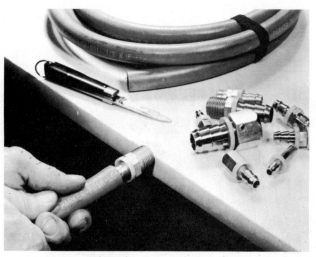

Courtesy Parker-Hannifin Corp.

Fig. 26. This hose assembly can be produced quickly with minimum tools.

CAUSES OF PIPING FAILURES

A number of piping troubles may develop. Any mistakes that are made in design and installation often result in costly maintenance and repair problems. Some of these causes are:

1. The operating pressure may be too high. Leaks and damage may take place.
2. The piping may not be properly supported. Either sudden surges or sudden stoppages of flow may cause banging or

Courtesy The Cross Company

Fig. 27. Hydraulically operated machine automatically assembles the piston-connecting rod assembly to the cylinder block-crankshaft subassembly. Note the well-designed piping layout. Also, note the various sizes of hydraulic tubing employed and how they are anchored at various points.

hammering in the piping. Banging of the piping may be experienced in some household systems.

3. The incorrect size of piping may cause trouble. A restric-

tion in the piping, as in an exhaust or return line, may cause a failure in the circuit.

4. Small piping may cause restrictions in pressure lines and reduce the functions of various devices.

5. Sometimes in a long run of piping, the pipe size reduces in the direction of flow and then increases. This is poor piping practice. Sometimes this type of piping results when additions are made to an existing system. As a general rule, oversize piping is usually a wise investment.

6. Copper tubing in a hydraulic system may fail because of work hardening of the copper.

7. In compressed air distribution systems, one major piping difficulty, called "low-pressure air," may occur. Normally, many pneumatic tools are designed to operate with an inlet pressure of about 90 pounds per square inch. If the air pressure entering the tool is too far below 90 *psi,* the tool cannot work properly. Possible causes of low-pressure air are: (1) insufficient compressor capacity; (2) inadequate piping; and (3) leakage. Again, it should be noted that oversize piping may be a wise investment. Inadequate piping may be the result of an addition to an original system. Frequently, leakage can be reduced to provide satisfactory air pressure.

REVIEW QUESTIONS

1. What classes of piping can be found in a home?

2. What classes of piping can be found in an automobile?

3. What classes of piping can be found in the place where you work?

4. What are some of the difficulties, that may arise in tubing installations, that can be avoided by installation of rigid piping?

5. What are some of the hazards of using flexible hose in an installation?

6. What are the causes of low-pressure air in a pneumatic distribution system?

7. What are some of the factors that should be checked before installing a piping system?

8. When installing pipe with threaded ends, what is the advantage of applying pipe compound or *Teflon* tape only on the male threads?

9. What difficulties may develop if a hacksaw is used to cut tubing?

10. What type of piping should be used from the intake filter to the pump on a hydraulic power device? Explain.

11. What class of pipe should be used for service that requires 200 *psi*? For service that requires 2000 *psi*?

CHAPTER 10

Air Filters, Pressure Regulators, and Lubricators

The air compressor and the various components that are located near the compressor were discussed in a previous chapter. The compressor system may include, in addition to the compressor, an aftercooler, a safety valve, a water separator, and a receiver. The compressed air leaving the compressor system is then distributed by means of piping to the various devices for doing work; these devices may be air cylinders, pneumatic motors, or pneumatic tools (Fig. 1). In a well-designed pneumatic system, the air must pass through an air filter, a pressure regulator, and a lubricator in the line, before it is delivered to the device or machine that performs the work. It is sometimes recommended that not more than two tools be served by a single filter, regulator, or lubricator. In this chapter, the functions, installation, and maintenance of the three components: (1) air filter; (2) pressure regulator; and (3) lubricator are discussed. All three components are necessary for efficient operation of pneumatic systems, and it is important that they should be installed and maintained properly.

GENERAL FEATURES

Some of the general operating features should be pointed out. Several different work devices or air tools may be connected to a single compressor, which may be located at a distance from the tools. There may be periods of time when no air is required from

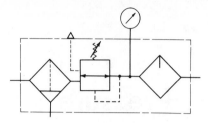

Fig. 1. Compressed-air circuits should consist of a filter, a regulator, and a lubricator.

the compressor. The distribution piping may be warm during some periods of time and cold at other intervals. It is almost impossible to keep the air distribution piping free from dirt, moisture or condensate, pipe compound, pipe scale, emulsified and deteriorated compressor oil, and rust.

Moisture in the air may condense, which enables rust to form, and pipe scale may peel. In different compressed-air plants, various amounts of impurities may be found in the air distribution lines. Foreign matter in the compressed air can do real damage, causing a system to become inoperative. Foreign matter can clog the valve ports and damage the valve parts and other close-fitting parts in air tools, air motors, pistons, and other devices. For example, pipe scale which consists largely of iron oxide is extremely abrasive; it can score the vanes in air motors and any other surfaces with which it comes in contact. Deteriorated oil from the compressor may be deposited on various surfaces, forming a sticky and gummy deposit that is sometimes called "varnish." It is essential that only clean air be supplied to the working devices; thus the air must be filtered.

To operate properly, it is necessary to maintain proper air pressure for each air device or tool. If a pneumatic tool requires a pressure of 90 pounds per square inch, gauge, the pressure at the compressor must be higher than the pressure at the individual tool. Pressure losses occur in the distribution piping, and work devices may be operated intermittently; thus the air pressure must be regulated to produce the proper volume at the point where it enters the work device or air tool.

The various moving mechanical parts in the different work devices—air cylinders, air motors, and pneumatic tools require proper lubrication for them to work efficiently, without undue wear. It is common knowledge that proper lubrication is essential for the proper operation of an automobile engine. Similarly, proper lubrication is essential for air-operated devices. The air filter, pressure regulator, and lubricator can be installed either in separate units or in a combined unit.

Air Filters

Ideally, the air filter should remove all foreign matter, and it should allow dry clean air to flow freely without resistance. A variety of air filters is available commercially. Both an inlet and an outlet, each with a female pipe thread, can be found in each filter housing. Marks on the housing indicate the direction of air flow, so that it can be installed correctly. A filter element or device for removing foreign matter can be found in the housing.

The filter should be installed in a line in such a way that it cannot be by-passed. If the filter is by-passed, the working devices may be damaged. It may be desirable to replace a filter with a stand-by filter while the filter is being cleaned and maintained, if shutdown time is to be minimized.

The filter should be large enough in capacity to handle the required flow of air. Various sizes of filters are available.

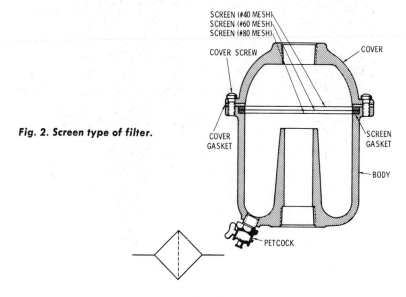

Fig. 2. Screen type of filter.

Various methods of filtering or removing foreign matter from the air are used, and various materials are used for the filter element. The air may be passed through a piece of porous metal, a porous stone, felt, resin-impregnated paper, or resin-impregnated wool fiber; or the foreign matter may be removed by a centrifuge or cyclone action.

The type of filter required depends on the size of the particles that are to be screened from the system. For example, some types of valves allow large particles to pass through them, without causing damage to the well-machined surfaces. As an illustration, a poppet-type valve (of the same type as that used on an automobile engine) may be able to accommodate larger particles than some other types of valves, such as the sliding-piston or spool-type valves.

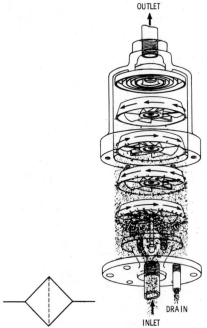

Fig. 3. Mechanical type of filter.

196

Frequently, the size of the foreign particles is designated in "microns." One micron is equivalent to 1/25,400 inch. Conversion values can be found in Table 1. Filters can be obtained that are capable of removing particles in each of the following ranges: under 5 microns; 5 to 9 microns; 10 to 24 microns; 25 to 49 microns; and over 50 microns.

In the screen type of filter (Fig. 2), air enters at the top, passes through the screen, and then passes out of the filter. The various parts are identified in the illustration.

In a mechanical type of filter (Fig. 3), air enters at the bottom. As the air passes through the housing, it rotates four rotors at a high speed. This action is similar to that of the centrifuge of a cream separator; the foreign particles, which are heavier than the air, are thrown against the outer walls of the housing by centrifugal force. The foreign particles then drop downward to the bottom of the housing and into a trap. Two rotors revolve in the same direction while the other two rotors revolve in the opposite direction. As the air stream passes from one rotor to another, the sudden reversal of the air stream provides a cleaning action for removing the foreign particles.

A type of filter with an element made of a phenolic-impregnated cellulose is diagrammed in Fig. 4; this material is electrically fused and polymerized for cohesiveness between layers, making the element impervious to gases, moisture, and common solvents. The plastic-impregnated material, in ribbon form, is wound edgewise on a mandrel to form a cylindrical element. This action is called "edge filtration," because the air passes through the ribbons. The air stream enters at the top of the housing, passes through the filter element, and then passes outward through the outlet. The foreign particles cling only to the outer surface of the element. The impurities can be removed by reversing the air stream.

Table 1. Micron-Inch Conversions

Microns	Inch
5	0.0002
10	0.0004
15	0.0006
20	0.0008
40	0.0016

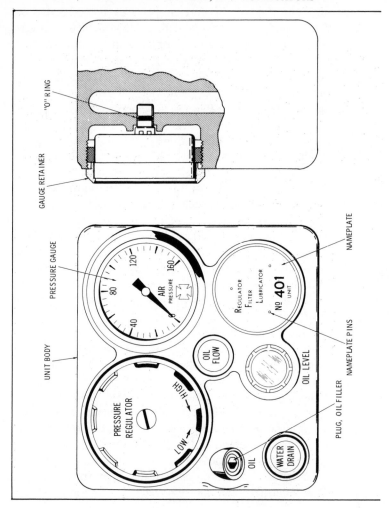

Fig. 4. A filter with a phenolic

Various combinations of edge and cyclonic filtration are possible (Fig. 5). Air enters at the left-hand port. A deflector plate provides a swirling or cylonic action to the air flow. The action is similar to that of a centrifuge; the larger foreign particles, which are heavier than the air, are thrown against the walls of the bowl

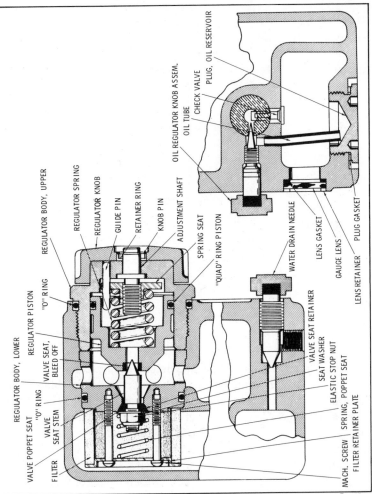

Courtesy Logansport Machine Co., Inc.

-impregnated cellulose element.

by centrifugal force. The smaller or finer particles are removed by the ribbon-type filter element.

An automatic-drain type of air filter is illustrated in Fig. 6. As the liquid accumulation level rises in the bowl, the float opens the pilot valve, admitting air to the pilot chamber. As the pressure

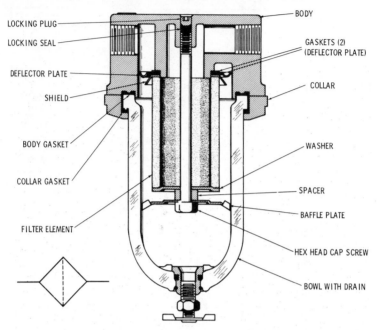

LOCKING PLUG

LOCKING SEAL

DEFLECTOR PLATE

SHIELD

BODY GASKET

COLLAR GASKET

FILTER ELEMENT

BODY

GASKETS (2)
(DEFLECTOR PLATE)

COLLAR

WASHER

SPACER

BAFFLE PLATE

HEX HEAD CAP SCREW

BOWL WITH DRAIN

Courtesy Parker-Hannifin Corp.

Fig. 5. Filter unit.

increases in the chamber, the diaphragm extends and opens the scavenger valve; thus liquid and other impurities are blown out the drain opening. Fig. 6 also provides charts that show pressure drop across the filter versus air flow (cubic feet per minute) through the filter. Fig. 7 provides information as to the correct sizes and dimensions of the various filters. Table 2 lists the commercially available types of air filters that can be used for pneumatic systems. The manufacturer's recommendations as to the type of filter should be followed.

Pressure Regulators

The pressure regulator is sometimes called a reducing, or regulating, valve. The air distribution line is connected to the inlet of the pressure regulator; the inlet pressure may be 140 to 150 pounds per square inch. The regulator provides a constant set pressure at the outlet of the regulator; this setting may range from "0" to full line pressure. Regulators may be either the diaphragm type or the piston type.

In a regulator of the diaphragm type (Fig. 8), the adjusting screw is used to set the valve for a given pressure at the reduced

Table 2. Listing Of Commercially Available Filters

Types	Bowl, canister, filler cap
Elements	Porous metal, porous stone, metal screen, paper, felt, cellulose, dessicant, reusable
Bowl	Metal, plastic, guarded
Capacity	0 to 1,000,000 cfm
Maximum pressure rating	20 to 20,000 psi
Minimum particle size rating	0.3 to 100 microns

side. After the valve has been set for a given pressure, the lock nut should be tightened to prevent a change of setting. The bronze diaphragm positions the disk, allowing the correct flow through the valve port that is required to maintain the desired pressure at the regulator outlet. The steel adjusting spring exerts a force on the upper side of the diaphragm. The reduced pressure on the outlet side affects the lower side of the diaphragm. A movement of the diaphragm also moves the disk, and changes the flow through the valve.

If a drop in pressure occurs on the discharge side of the regulator, the upward force on the diaphragm is decreased. Thus, the balance is disturbed, and the adjusting spring pushes the diaphragm downward. This action moves the disk to increase the size of the opening of the valve port, thereby increasing the flow through the valve.

If an increase in pressure occurs on the discharge side of the regulator, the upward force on the diaphragm is increased. The balance is disturbed, and the diaphragm is pushed upward. This action moves the disk to close the valve, thereby reducing the flow through the valve.

In installing a regulator, care should be taken to avoid using an excess of pipe joint compound, as pipe compound may block the mechanism. This difficulty can be avoided by placing the proper quantity of pipe compound on only the male threads.

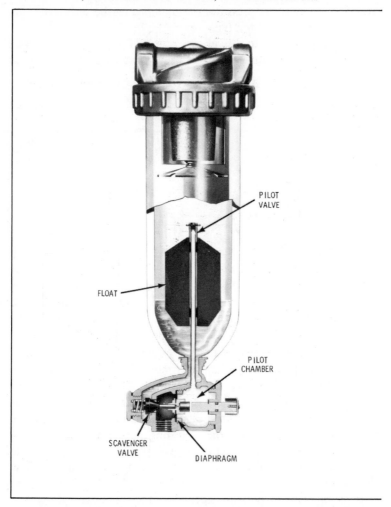

PILOT
VALVE

FLOAT

PILOT
CHAMBER

SCAVENGER
VALVE

DIAPHRAGM

Fig. 6. Automatic-drain type of air filter with a 40-micron filter element,

A piston-type regulator is shown in Fig. 9. The spring acts on the top side of the piston which consists of the piston cup and a metal backing disk. The force of the spring can be changed by means of the adjusting screw. The regulator body acts as a cylinder for movement of the piston. The inlet air enters the chamber, exerting a force on the lower side of the piston; it passes

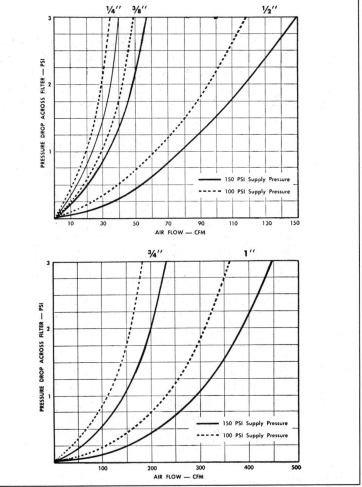

Courtesy Watts Regulator Company

and charts that indicate pressure drop across the filter versus air flow.

over the valve seat and outward through the outlet port. The valve seat is supported by the yoke, and is guided by the seating guide on the lower cap. If the pressure on the inlet side of the valve is decreased, the force of the spring increases the size of the valve opening, allowing a larger flow through the valve.

In setting a regulator, a pressure gauge is convenient on the

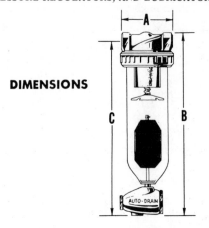

DIMENSIONS

Size		Bowl		Dimensions (in.)			Wt.
(in.)	No.	Capacity	Material	A	B	C	Lbs.
¼, ⅜, ½	D602	1 pt.	Plastic	3½	11⅞	11¼	3⅞
¼, ⅜, ½	D603	1 pt.	Metal	3½	11⅛	10½	4⅞
¾, 1	D602	1 qt.	Plastic	4⅝	13¾	12⅞	5¾
¾, 1	D603	1 qt.	Metal	4⅝	12⅞	12	6⅜

Courtesy Watts Regulator Company

Fig. 7. Dimensions and weights of the various sizes of automatic-drain type of filters.

discharge side of the regulator; this can be used to indicate when the pressure is adjusted properly.

The performance characteristics of a typical air pressure regulator are shown in Fig. 10. This type of information can be furnished by a manufacturer to aid in selecting a regulator and in determining its performance in a given system.

Lubricators

Different types of air line lubricators are found in actual practice. One important feature of many lubricators should be explained before discussing the basic action of a lubricator. As illustrated in Fig. 11, fluid flowing in a pipe reaches the converging section, where the pressure is designated P_1. As the pipe converges, the converging section acts as a nozzle. The fluid velocity at the end of the converging section is higher than the velocity at the beginning of the section. This action is similar to that in

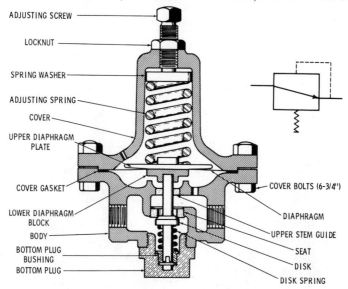

ADJUSTING SCREW
LOCKNUT
SPRING WASHER
ADJUSTING SPRING
COVER
UPPER DIAPHRAGM PLATE
COVER GASKET
LOWER DIAPHRAGM BLOCK
BODY
BOTTOM PLUG BUSHING
BOTTOM PLUG

COVER BOLTS (6-3/4")
DIAPHRAGM
UPPER STEM GUIDE
SEAT
DISK
DISK SPRING

Courtesy The Clark-Reliance Corp.

Fig. 8. A diaphragm-type pressure regulator.

the nozzle of a garden hose or a fire hose. The pressure P_2 is lower than the pressure P_1. A decrease in pressure is associated with the increase in fluid velocity. Then, as the channel diverges, the fluid velocity downstream becomes higher than that at the converged section. This arrangement of a flow channel is often called a "venturi" after the scientist, Venturi. A similar action can be developed by means of an orifice, or a hole in a plate. A "venturi passage" is used in automotive carburetors to draw gasoline into an air stream. A similar arrangement is used in the air line lubricators to draw oil into the air stream.

All moving mechanical parts require lubrication. Usually, the savings from reduced repair costs and increased efficiency are greater than the cost of installing the lubricating equipment. Generally speaking, pneumatic components are lubricated by means of a lubricator placed in the air line; the lubricator injects oil into the air stream which is then delivered to the component.

The lubricator serves two purposes: (1) it stores oil; and (2) it injects a metered or adjusted quantity of oil mist, or oil fog, into

205

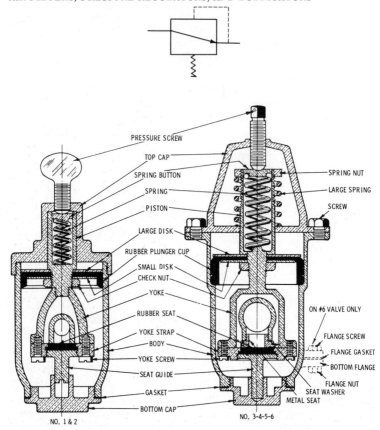

Fig. 9. A piston-type pressure regulator.

the moving air stream. The oil may be stored in either a metal or a plastic bowl, depending on the surrounding temperature and the air pressure. Metal bowls are usually recommended for pressures over 150 psi and for installations located near high-temperature areas, such as furnace areas.

A typical air line lubricator is diagrammed in Fig. 12. The transparent bowl is filled with oil; the oil level can always be seen. Air enters at the left-hand side, passes through a venturi passage, and leaves at the outlet on the right-hand side. The reduced pressure at the middle section of the venturi helps to draw the oil from the bowl into the air stream. The ball check valve in the feed tube maintains oil in the tube at all times. A vane assembly in the middle section of the venturi can be adjusted to control the amount of oil that is drawn into the air stream. Oil reaches the air stream in the form of extremely fine particles, and is carried suspended in the form of a mist in the air stream.

Table 3. Air Flow Rates For Oil-Fog Lubricators

(cfm = cubic feet per minute, free air)

Operating Pressure (psi)	1/4" Pipe cfm		3/8" Pipe cfm		1/2" Pipe cfm		3/4" Pipe cfm		1" Pipe cfm		1 1/2" Pipe cfm	
	Min	Max	Min	Max	Min	Max	Min	Max	Min	Max	Min	Max
10	2	10	5	25	11	55	23	115	43	215	103	515
20	2	10	6	30	13	65	27	135	51	255	122	610
30	3	15	7	35	14	70	32	160	58	290	138	690
40	3	15	7	35	16	80	35	175	64	320	153	765
50	3	15	8	40	17	85	38	190	69	345	166	830
60	4	20	9	45	19	95	42	210	75	375	179	895
70	4	20	10	50	20	100	45	225	79	395	190	950
80	4	20	10	50	21	105	47	235	84	420	202	1010
90	4	20	11	55	22	110	50	250	88	440	212	1060
100	4	20	11	55	23	115	52	260	92	460	222	1110
110	4	20	12	60	24	120	54	270	96	480	232	1160
120	5	25	12	60	25	125	56	280	100	500	242	1210
130	5	25	12	60	26	130	58	290	104	520	252	1260
140	5	25	13	65	27	135	60	300	108	540	261	1305
150	5	25	13	65	28	140	62	310	111	555	269	1345

Courtesy C. A. Norgren Co.

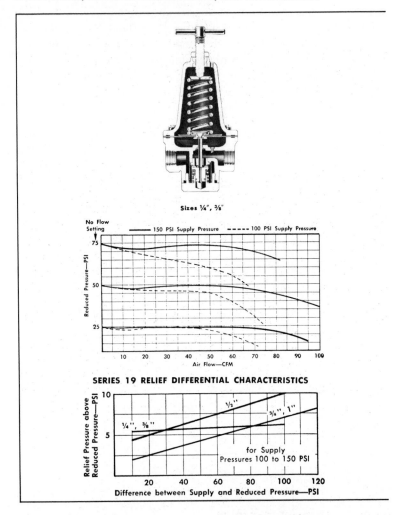

Fig. 10. Performance characteristics

Many lubricators make use of a venturi, or orifice, to develop a differential pressure; thus the oil is lifted or siphoned from the bowl. A needle valve can be used in the oil tube to adjust the flow of oil into the air stream. In a wick-type lubricator, a wick transfers the oil from the bowl to the air stream. The action of a wick-fed lubricator can be adjusted by either increasing or decreasing

CAPACITY AND
QUICK SIZING CHART
SERIES 18, 19

Size Inches	Supply Pressure	CFM Air Flow at 10% Drop from Reduced Pressure Setting of:		
		25	50	75
¼	100	45	25	20
	150	60	50	35
⅜	100	50	55	25
	150	75	80	60
½	100	120	40	30
	150	185	130	50
¾	100	240	230	90
	150	350	345	150
1	100	260	230	100
	150	375	360	165

EXAMPLE: What size regulator must be used to deliver 125 CFM air at 50 PSI from a supply pressure of 150 PSI? Under the "50" column and opposite a "150" it is found that the ½" size will deliver 130 CFM.

SERIES 19 RELIEF CAPACITY—CFM

Size Inches	Relief Pressure Above Reduced Pressure PSI	Reduced Pressure—PSI					
		10	25	50	75	100	125
¼, ⅜, ½	5	1.2	1.2	1.2	1.2	1.4	1.5
¾, 1		—	—	—	—	2.0	4.2
¼, ⅜, ½	10	3.3	4.0	4.6	4.8	4.9	5.2
¾, 1		4.0	4.5	4.5	5.0	6.5	8.0
¼, ⅜, ½	15	4.5	5.4	6.2	6.6	6.8	7.2
¾, 1		7.5	8.0	8.0	8.0	9.2	9.6
¼, ⅜, ½	20	5.5	6.4	7.3	7.8	8.0	8.5
¾, 1		10.2	10.5	10.5	10.5	11.4	11.5
¼, ⅜, ½	25	6.2	7.2	8.1	8.8	9.0	9.5
¾, 1		—	—	—	—	—	—

EXAMPLE: A ½" Series 19 Regulator is operating a cylinder at 50 PSI, if pressure is built-up to 60 PSI how much air will be released? Under the 50 PSI column and opposite "10" for the ½" size find 4.6 CFM.

INVERSE EFFECT OF SUPPLY PRESSURE VARIATION on
REDUCED PRESSURE

SERIES 18, 19

Supply Pressure Variance PSI		+50	+25	+15	+10	+5	0	−5	−10	−15	−25	−50
Reduced Pressure Change PSI	¼", ⅜"	−1.1	−.6	−.3	−.2	−.1	0	+.1	+.2	+.3	+.6	+1.1
	½"	−2.6	−1.3	−.8	−.5	−.3	0	+.3	+.5	+.8	+1.3	+2.6
	¾", 1"	−3.0	−1.5	−.9	−.6	−.3	0	+.3	+.6	+.9	+1.5	+3.0

of an air pressure regulator.

the length of wick extending into the air stream. A minimum air velocity is required for proper lubrication of air lubricators. The recommendations of one manufacturer are given in Table 3. Line pressure and air flow are important considerations in the size of a lubricator. The table follows the fundamental rule that, for efficient operation, the maximum flow rate should not exceed five

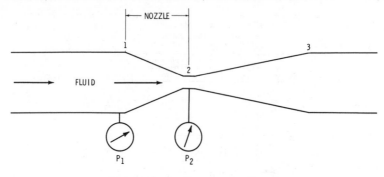

Fig. 11. Venturi action in a lubricator.

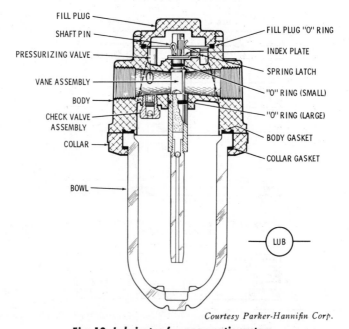

Courtesy Parker-Hannifin Corp.

Fig. 12. Lubricator for pneumatic system.

times the minimum flow rate. Table 3 shows maximum and minimum flows at the various operating pressures. For example, if an operating pressure of 80 psi and a flow-rate of 70 cfm are required, Table 3 indicates that either a 1/2-inch or a 3/4-inch lubricator can operate at 70 cfm.

It is important that the equipment should receive some oil continuously. Lack of oil causes wear. Too much oil may result in flooding and sluggish operation.

For many applications, a top grade of spindle oil provides suitable lubrication. Another lubricant which has been used successfully consists of a mixture of 50 percent kerosene and 50 percent SAE 30 oil. Manufacturers and competent oil dealers should be consulted for recommendations for specific machines and equipment. Many lubricator manufacturers recommend a petroleum-base oil with a Saybolt Universal Reading of 80 to 150 seconds at 100°F. Lubricators can handle a more viscous oil, but such an oil may cling to the inside of the lines and thus resist movement by the air flow. At extremely low temperatures, it may be necessary to heat the oil. Lubricators with the heating elements controlled by a thermostat are available.

Lubricators may be filled either manually or automatically. The lubricator bowl is placed under air pressure during operation. Thus, the air pressure to some lubricators must be shut off before removing the filler plug. In some lubricators, a check valve may stop the air flow into the bowl when the filler plug is removed. A portable oil pump can be used to fill a lubricator that is under pressure.

Some maintenance is necessary to maintain proper oil delivery. The condensate should be drained from the lubricator bowl, and the bowls cleaned with either kerosene or soap and water.

Lubricators should be installed as close as possible to the equipment that is being lubricated. They should be placed downstream from the air filter and the regulator; this provides clean, regulated air, and reduces the restrictions between the lubricator and the part that is being lubricated.

Information pertaining to commercially available lubricators can be found in Table 4.

Combination Units

Units consisting of a filter, regulator, and lubricator are frequently connected together in a pneumatic system. Three units connected together with close nipples are shown in Fig. 13, and

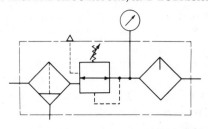

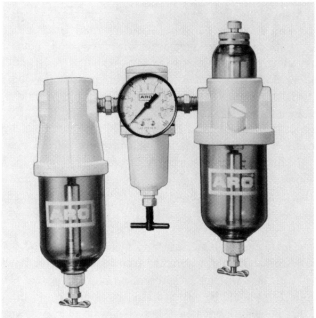

Courtesy The ARO Corporation

Fig. 13. Filter, regulator, and lubricator units in a pneumatic system.

Table 4. Commercially Available Air Lubricators

Oil-feed type	Drop, wick, siphon, mist or fog
Bowl	Metal, plastic, reinforced plastic, metal guard
Filling	Manual, automatic, automatic air shutoff
Oil feed control	Adjustable, fixed, self-adjusting
Maximum oil capacity	6 to 640 ounces
Minimum air flow to lubricate	0.1 to 10 cfm
Maximum pressure rating	125 to 1000 psi

the three components can be built into a single housing or combination unit (Fig. 14).

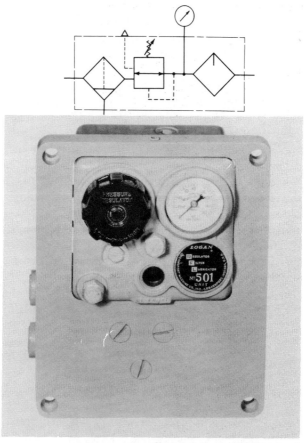

Courtesy Logansport Machine Co., Inc.

Fig. 14. A combination unit in which filter, regulator, and lubricator are built into a single housing. Panel also has a four-way pneumatic control valve.

PROTECTION OF FILTER AND LUBRICATOR BOWLS

A transparent bowl is advantageous in determining visually when oil should be added to a lubricator and when an air filter should be drained. The breakage of a bowl on an air filter or

213

air lubricator is a possible hazard. Precautions are necessary to reduce this hazard.

Improper application or poor installation practices may result in bowl breakage. For example, a bowl should not be placed where it may be broken by a passing vehicle. A plastic bowl may not be suitable in an area where the temperature is high, or in a chemical atmosphere which may deteriorate the plastic material. Glass may be considered as a bowl material in some installations, but it may be more expensive.

Metal bowls are usually made of brass, steel, or aluminum, and are usually recommended for operations in: (1) regions where the temperature may rise above 120°F., such as areas near furnaces, compressors, and forging presses; (2) regions of high or violent vibration; (3) regions of high pressure; and (4) regions where the atmosphere contains chemical fumes or solutions which attack plastic bowls.

Various designs of bowl guards which provide both visibility and protection are available. In one design, a perforated metal cover fits over the plastic bowl; thus the bowl is protected against impact, and the pieces are restrained if the bowl should break. In another design, a coiled steel wire is attached to the piping by means of hooks. In other designs, wire mesh is molded into the plastic bowl, and a pyrex window is placed in the metal bowl.

MUFFLERS

The exhausting of compressed air from control valves and other pneumatic devices can be quite annoying, especially if there are a number of the devices in a relatively small confined area. Much emphasis is placed on reducing noise levels in industry by the government, labor unions, insurance companies, and others. Much has been accomplished during the past decade to alleviate noise problems. Pneumatic mufflers of various types are employed with excellent results. One of these mufflers is shown in Fig. 15. The exhaust air enters the muffler (see Fig. 15B) at a high velocity. The exhaust air mass is redirected and subdivided into separate infinitely smaller air streams. Directed in a continuous stream at the wall of the obstruction-free expansion chamber, the smaller air streams rebound to collide head-on with opposing air streams of

A

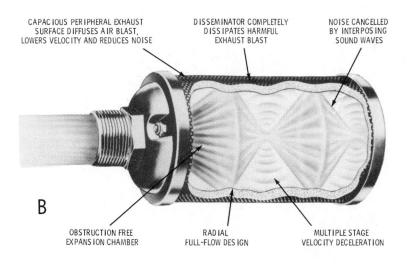

CAPACIOUS PERIPHERAL EXHAUST
SURFACE DIFFUSES AIR BLAST,
LOWERS VELOCITY AND REDUCES NOISE

DISSEMINATOR COMPLETELY
DISSIPATES HARMFUL
EXHAUST BLAST

NOISE CANCELLED
BY INTERPOSING
SOUND WAVES

B

OBSTRUCTION FREE
EXPANSION CHAMBER

RADIAL
FULL-FLOW DESIGN

MULTIPLE STAGE
VELOCITY DECELERATION

Courtesy Allied Witan Company

**Fig. 15. Pneumatic muffler. (A) Note pipe connection and disseminator.
(B) Cross section showing flow path of the exhaust air.**

equal force at a point predetermined by the design of the muffler.
At a greatly reduced velocity, the air is then dispersed through the
openings in the increased area of the disseminator surface provided
by the perforated cylinder walls of the muffler at a very low noise
level.

The mufflers are manufactured in pipe port sizes from ⅛ to 6
inches and sometimes larger. Also, these mufflers play an impor-

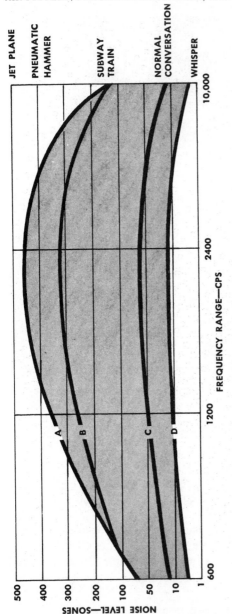

FACTS ABOUT NOISE

Curve A—magnitude of loudness when exhaust air is discharged directly into atmosphere without a muffler.

Curve B—Exposure to noise above this level is difficult to endure, hearing loss can result from continuous exposure, protection required.

Curve C—Exposure to noise above this level causes fatigue, poor efficiency and costly errors, protection recommended.

Curve D—attenuated level when ATOMUFFLER is attached to the exhaust opening of the equipment.

Shaded area illustrates the effective elimination of air exhaust noise with an ATOMUFFLER used on air-operated equipment.

This graph was prepared to show exhaust noise loudness in sones ranging from loud to soft as distinguished by the human ear. In defining loudness sones differ from decibels in that a decibel is a numerical value indicating intensity of noise, whereas sones indicate the actual loudness as heard by the human ear.

Courtesy Allied Witan Company

Fig. 16. Facts about noise.

tant role in a pneumatic system, and they should be properly maintained and cleaned.

The facts about noise are given in Fig. 16. The noise level, in sones, indicates the actual loudness heard by the human ear.

A high-production assembly machine utilizing pneumatic equipment to perform the required movements is shown in Fig. 17.

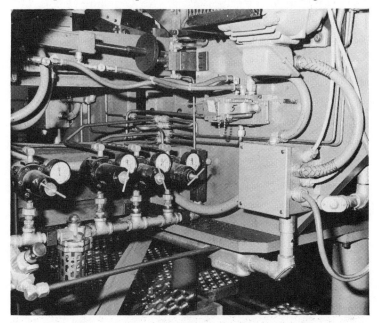

Courtesy Gilman Engineering & Manufacturing Co.

Fig. 17. This high-production assembly machine utilizes pneumatic equipment to perform the required movements. Note the four pressure regulators and the lubricator in the foreground.

REVIEW QUESTIONS

1. What are the disadvantages of by-passing the filter in a system?

2. What is the difference between cyclonic and edge filtration?

3. Why is a pressure regulator needed in the system?

4. Explain the reasons for placing a lubricator in the air line.

5. Why is a pressure gauge placed in an air distribution system?

6. What are the disadvantages of using a single air compressor to serve a large number of different air tools?

7. Draw a schematic digaram, using *ANS* Graphic Symbols, of a filter with manual drain, a regulator and a lubricator all in one housing; a shutoff valve upstream from this unit; and a regulating valve and gauge on a tee-line downstream from the packaged unit.

CHAPTER 11

Directional Control Valves

The function of a directional control is to direct oil or air to various places in the system. The directional control directs the movement of the fluid so that it can do work. The directional control is a valve. Three different classes of directional control valves are used in fluid systems. They may be listed as: (1) two-way (two ports and two internal passages); (2) three-way (three ports and three internal passages); and (3) four-way (four ports and four internal passages). Each valve may be operated manually, mechanically, electrically, or by a pilot arrangement.

In this chapter, the function, installation, and maintenance of directional control valves are to be studied. The importance of proper operation of directional controls cannot be overemphasized. How well a system functions often depends on the selection of the proper control. The two-way valve is discussed first, and then the three-way and four-way valves are discussed in order.

TWO-WAY VALVES

A two-way valve consists of two ports—a port which lets the fluid come in (inlet port), and a port which lets the fluid come out (outlet port). Each port is connected to an internal passage. At some point in the operation of these valves, the two passages are connected; at another point, they are disconnected. Two-way valves are either "normally open" or "normally closed." This means that in the normally-closed valve, when the actuator (the mechanism which makes the valve function) is in its normal position, the inlet port and passage are shut off from the outlet port and passage. To connect the two passages, the valve operator must function. In a normally-open valve, the inlet passage and the

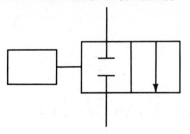

Fig. 1. ANS Graphic Symbol for a
two-way control valve.

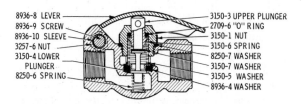

8936-8 LEVER
8936-9 SCREW
8936-10 SLEEVE
3257-6 NUT
3150-4 LOWER
    PLUNGER
8250-6 SPRING

3150-3 UPPER PLUNGER
2709-6 "O" RING
3150-1 NUT
3150-6 SPRING
8250-7 WASHER
3150-7 WASHER
3150-5 WASHER
8936-4 WASHER

Courtesy Scovill Fluid Power Division, Scovill Mfg. Co.

Fig. 2. A poppet-type two-way valve.

outlet passage are connected when the actuating mechanism is in
normal position. The valve actuator must function to separate the
passages. Two-way valves have many uses. Some of these are: to
shut off air or oil flow, to bleed off pressures, and to actuate pilot-

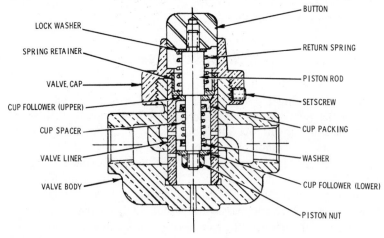

LOCK WASHER
SPRING RETAINER
VALVE CAP
CUP FOLLOWER (UPPER)
CUP SPACER
VALVE LINER
VALVE BODY

BUTTON
RETURN SPRING
PISTON ROD
SETSCREW
CUP PACKING
WASHER
CUP FOLLOWER (LOWER)
PISTON NUT

Fig. 3. Diagram of a two-way valve.

operated control valves. The *ANS* Graphic Symbol for a two-way valve is shown in Fig. 1.

Manual Control

In a manual, or by man, control, the means of moving the valve operating mechanism is by hand, by foot, or by some part of the human body, and some human force is required to perform this act. Manual controls are seldom used in automatic operation, but they perform very important functions in our industrial processes. Two-way hydraulic valves are built in pressure ranges that may exceed 10,000 psi, but most applications are under 1500 psi. Pneumatic two-way valves usually do not exceed 150 psi. In industrial fluid power applications, the port sizes in the majority of applications are under 2-in. pipe size.

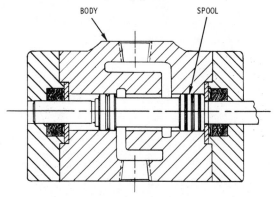

Fig. 4. Illustrating metal spool in a two-way valve.

Manual Operation

The names of the most important parts which make up a two-way valve should become familiar. It is nearly impossible to cover all designs of two-way valves, or any other type of control valve. New designs are constantly being devised. However, there are certain basic parts which should be learned. Two-way valves are made in the following designs: poppet (Fig. 2); sliding spool (Figs. 3 and 4); diaphragm (Fig. 5); plug (Fig. 6); and disk (Fig. 7). Many of these valves are in common use.

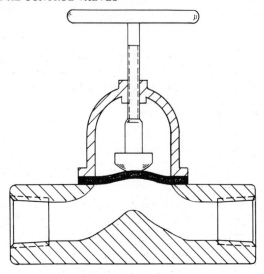

Fig. 5. Diaphragm-type two-way valve.

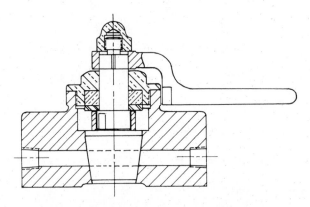

Fig. 6. Plug-type two-way air control valve.

Valve Body—The valve body is usually made of iron, aluminum, or bronze. The body must be pressure tight, or fluid under pressure can seep out. The valve body consists of two pipe ports and some type of mounting feet, or pads, unless it is a type that uses the pipe ports for mounting. If it is a manifold-mounted valve, the pipe ports are located in a separate plate, and the valve body is fastened to the plate with mounting bolts. The chief advantage of a valve body that is mounted on a manifold plate is that the valve can be removed without disturbing the pipe. Some valves use liners inside the body.

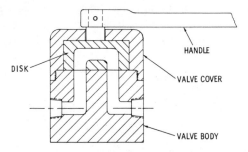

Fig. 7. Disk-type two-way valve.

Valve Cover—The valve cover is usually a casting or a screw machine part which is generally of the same material as the valve body. The cover may be fastened to the valve body with screws, or it may be threaded into the valve body. The valve cover usually helps to hold the working parts in place within the valve body.

The Actuator—The actuator is the device which operates the valve mechanism. In a manually operated valve, the actuator may be a foot pedal, a hand lever, a push button, or even a toggle. The actuator is usually made of a material that can withstand much abuse.

Valving Mechanism—The valve or valving mechanism depends on the type of valve. The valving mechanism in Fig. 3 is composed of a piston with two cup packings. In Fig. 4, an alloy-steel spool is shown. A poppet-type valve uses poppets as the valving mechanism. The plug-type valve uses a tapered plug, and the diaphragm-type valve uses a flat diaphragm.

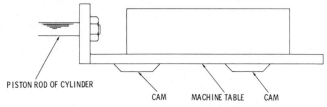

Fig. 8. Cams on a machine table.

223

In Fig. 3, when the actuator is in the normal position, the cup packing on the piston blocks the inlet passage from the outlet passage. When the actuator is operated, the piston moves so that the lower set of holes in the valve liner, which is connected to the outlet passage, is uncovered; and the fluid flows outward through the outlet port. In Fig. 4, the actuation occurs in a similar manner.

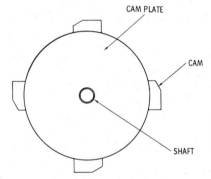

Fig. 9. Rotary cam arrangement.

Mechanical Operation

Directional controls can be operated by a number of mechanical means, such as a cam on a machine member (Fig. 8), a rotary multiple-cam arrangement (Fig. 9), a swing-type cam (Fig. 10), or a trip pin (Fig. 11). Other ways of actuating a mechanically operated valve are possible. The function of mechanically operated controls is to lend itself to application in a semiautomatic- or automatic-cycle circuit. With so much stress being placed on automation, industry is finding many more applications for

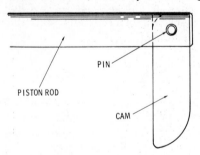

Fig. 10. Swing-type cam arrangement for actuating a valve.

mechanically operated valves. Whether these mechanically operated valves are of two-, three-, or four-way design, several important types of actuators are in use; some of these types are direct cam roller, offset cam roller, locking toggle, spring offset toggle, and button. The different types of actuators should be studied to become familiar with them, as they will be seen many

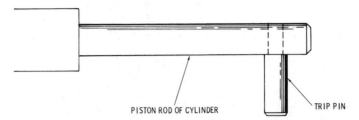

PISTON ROD OF CYLINDER   TRIP PIN

Fig. 11. Trip pin for actuating valve mechanism.

times in working with oil and air valves. In the preceding discussion, manual controls which the operator moves only so far to actuate the valve mechanism have been studied. If he tries to move them any further, he normally does not have the strength to cause damage to the valve. In a mechanically operated valve, however, the actuator is operated by some type of mechanism, which if not correctly designed, can cause considerable damage.

The following terms should be learned:

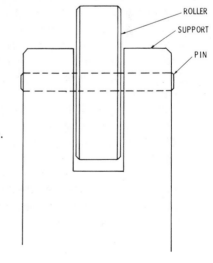

ROLLER

SUPPORT

PIN

Fig. 12. Cam roller mechanism.

225

1. *Cam roller.* A cam roller is a hardened wheel on a cam-operated valve which bears the force of the cam and is depressed by the cam to actuate the valve. The cam roller is pinned to the stem of the valve in the direct-acting cam roller type. The cam roller must be free to rotate, so that no galling action occurs between the cam and the cam roller and no flat spot is worn on the roller (Fig. 12).

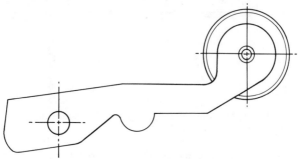

Fig. 13. Cam roller arm.

2. *Cam roller arm.* A cam roller arm is used on a valve with an offset arrangement. This arm provides a mechanical advantage which makes the roller more easily depressed. The length of the arm can be designed to provide the mechanical advantage that is required. A lobe on the cam arm acts against the actuator as the roller is depressed (Fig. 13).

3. *Toggle.* A toggle is a lever arm that can be actuated by some tripping means which, in turn, actuates the valve mechanism. Toggles are usually made of a hardened steel to resist wear and impact (Fig. 14); they are frequently used on control valves.

4. *Locking toggle.* A locking toggle is similar in design to the above toggle, except that when the lever arm is moved to its far position, it does not automatically return to its normal position when it is released. It must be actuated to be returned (Fig. 15).

5. *Pin actuator.* A pin-type actuator is a small button-type actuator that may be actuated by a pin or other tripping means. It is usually returned to its original position by spring pressure (Fig. 16).

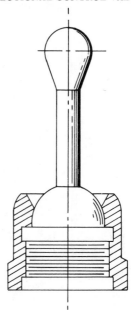

Fig. 14. Toggle lever.

Electrical Operation

Unlike manually and mechanically operated controls, only a limited amount of power is available to actuate the valve mechanism of electrically operated valves, and this power must be used to the best possible advantage. The operating means in an electrically operated valve may be an electric motor or a solenoid. An electric motor is seldom used in industrial pneumatics or hydraulics as an operating means; but when there is a large passage of low-pressure fluid involved, such as through a butterfly-type valve (Fig. 17), electric motors are used to advantage. Electric motors usually operate through gears and gear reducers to actuate the butterfly. These motors are usually reversible.

Solenoids are often used to operate valves. A solenoid is an electrical device that converts electrical energy into straight-line motion and force, as shown in Fig. 18. A solenoid consists of a coil of wire mounted on a soft-iron spool. The force developed is different at various points along the solenoid plunger travel. The force developed is increased as the solenoid plunger reaches the end of the plunger travel on the pull-type solenoid. The travel of a

227

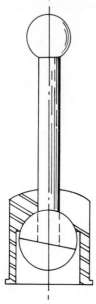

Fig. 15. Locking toggle.

solenoid plunger is rather short. An idea of the amount of force developed by a pull-type solenoid at various points along its plunger travel, may be obtained from the following figures: 1-in. stroke, 26 pounds; 7/8-in. stroke, 30 pounds; 1/2-in. stroke, 31 pounds; 1/4-in. stroke, 33 pounds; and 1/8-in. stroke, 40 pounds.

Solenoids are also used to operate a mechanical operator which, in turn, operates the valve mechanism. As indicated in Fig. 19, solenoids may be connected directly to the valve mechanism. Solenoids may be of the push type, pull type, or the push-pull type. The push-type solenoid is one in which the plunger is "pushed" when the solenoid is energized with electricity, and the pull-type solenoid is one in which the plunger is "pulled" when the solenoid is energized. The push-pull type of solenoid uses two coils; and the plunger is either pushed or pulled, depending on which coil is energized.

The names of the parts of the solenoid should be learned, so that they can be recognized when called upon to make repairs, to do service work, or to install them. The important parts are:

1. *Coil.* The solenoid coil is made of copper wire. The layers

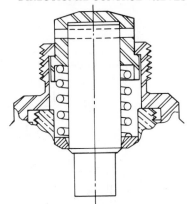

Fig. 16. Pin-type actuator.

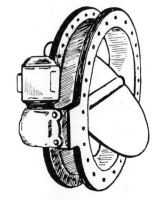

Fig. 17. Butterfly-type valve with motor control.

of wire are separated by insulating paper. The entire solenoid coil is covered with a varnish that is not affected by
solvents, moisture, cutting oil, or other fluids. Coils are rated
in various voltages, such as: 115 volts AC, 230 volts AC,
460 volts AC, 575 volts AC, 6 volts DC, 12 volts DC, 24
volts DC, 115 volts DC, and 230 volts DC. They are designed for such frequencies as 60, 50, and 25 cycles per
second.

2. *Frame.* The solenoid frame serves several purposes: (a)
Since it is made of laminated steel sheets, it is magnetized
when the current passes through the coils; the magnetized
frame attracts the metal plunger and causes the plunger to

229

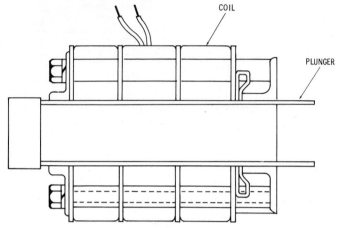

Fig. 18. Sketch of a solenoid.

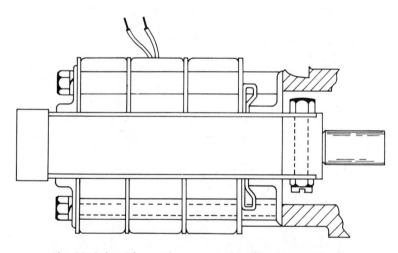

Fig. 19. Solenoids may be connected to the valve mechanism.

move. (b) The frame has provisions for attaching the mountings. Mountings are usually bolted or welded to the frame. (c) The frame has provisions for receiving the plunger. The wear strips are mounted to the solenoid frame, and are made of such materials as metal or impregnated glass fiber cloth.

3. *Solenoid plunger.* The solenoid plunger is the moving mechanism of the solenoid. The plunger is made of steel laminations which are riveted together under high pressure, so that there will be no movement of the laminations with respect to one another. At the top of the plunger, a pin hole is placed for making a connection to some device. The solenoid plunger is moved by a magnetic force in one direction, and is usually returned by spring action.

Solenoid-operated valves are usually provided with a cover over either the solenoid or the entire valve. This protects the solenoid from dirt and other foreign matter, and protects the actuator. In many applications, it is necessary to use explosion-proof solenoids. Underwriter's list for Class 1, Group D for hazardous locations (such as atmospheres which may contain gasoline, petroleum, alcohol and other highly inflammable fluids) requires solenoids with explosion-proof covers.

Most of the two-way solenoid valves are of the smaller types, and the solenoid plunger is connected directly to, or is a part of, the valve operating mechanism, as illustrated in Fig. 20. Frequently, these valves are used for operation of pilot-operated controls. They are small, compact, and can be cycled at a high rate, which is necessary in mass production. It is not uncommon to cycle more than one hundred times per minute.

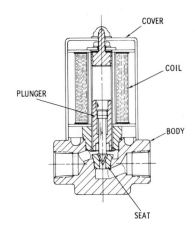

Fig. 20. Two-way solenoid valve.

231

Pilot Control

In this discussion, directional control valves that are operated by pilot controls are studied. This type of valve is very important in the functioning of automatic machinery. Pilot-operated controls are made in two-, three-, and four-way types. These valves may be operated by either two- or three-way pilot valves, depending on the type involved. Fig. 21 illustrates how a pilot-operated valve is operated in a system. The main valve (two-way) may be a large valve, as large as a four- or six-inch pipe size, and it may be located at a considerable distance from the small manually operated three-way valve. It may be placed at such a distance that the operator cannot see the valve. The valve at the operator's station may be as small as 1/4-inch pipe size. By merely shifting the handle of the small valve, which may require only a few ounces of effort, the operator can control the action of the large valve.

The advantages of a pilot-operated valve are:

1. Less effort is required on the part of the operator to control the valve mechanism of large valves. If the valve were

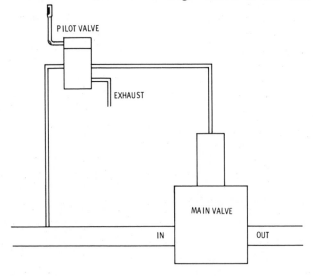

Fig. 21. Pilot valve operates the main valve.

operated by direct manual means, considerable effort would be expended in operating the valve mechanism.

2. The valve may be located at the scene of action, but the operator can be located at a vantage point where he can be in charge of a number of controls. The operating valve may be necessarily located in a hazardous area, such as near chemical fumes, explosive atmospheres, or in adverse temperatures. Yet, the operator can be spared all these discomforts, since the small valve which he operates may be located in a separate room or area.

3. The pilot-operated valve is not subject to requirements placed on electrically operated valves, since it is not necessary to have electric current present.

4. Since the pilot-operated valves can be placed near the devices that they are actuating, long runs of large piping can be eliminated, resulting in considerable savings. This saves not only on the piping but also on the installation. Large-diameter pipe is more difficult to install than small-diameter pipe.

5. When pilot-operated valves of the four-way type are used, the use of short piping makes it easier to exhaust or drain the fluid from the lines.

It is important to know how these valves function. As illustrated in Fig. 22, the fluid pressure from the small three-way valve moves

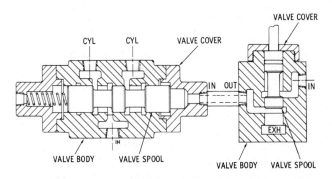

Fig. 22. Four-way pilot-operated valve (left), and three-way pilot valve (right).

through the pipe line to the pilot chamber of the four-way valve. The pilot chamber may be constructed as shown. The main part of the valve may be similar in design to other types of valves. The pilot-operating mechanism is used in place of a manual, mechanical, or electrical operation.

The pilot chamber is usually a casting or a screw machine part. It may have holes for mounting to the valve body; or it may be threaded, so that it can be screwed to the valve body. A pipe port is located at the end opposite the end that attaches to the valve body. This pipe port is usually made for 1/4-inch pipe connection. If the chamber contains a piston, the walls of the chamber must be very smooth in order to make its action as free as possible. The length of the chamber depends on the size of the operating mechanism, plus the distance required for the mechanism to move. The bore of the chamber depends on the amount of force required to move the valve operating mechanism when full pressure (plus a safety factor) is applied to the valve.

The pilot actuator may be in the form of either a piston spool or a diaphragm. As the fluid pressure contracts the actuator, it moves the valve mechanism and causes action within the valve. If the actuator is in the form of a piston, it will probably have a seal built into it; this seal may be in the form of an "0" ring, cup packing, a quad-ring, or some other commonly used packing. If a diaphragm is used, it is probably made of a synthetic rubber material.

Installation

The use of good judgment and a few simple rules aid in the installation of two-way valves.

Manual Operation—Here are some of the important rules for installation of two-way valves:

1. If the valve has mounting feet, mount them securely on a flat surface. Do not throw a strain on the valve body by mounting the valve on an uneven surface.
2. Install the valve so that the actuator is readily accessible to the operator. Do not require him to reach up and over a moving mechanism to get at the valve actuator.

3. Install the valve so that it is not subject to hot blasts. Extreme heat has a detrimental effect on most packings.

4. Install the valve so that it is not covered with dirt. Valves often become covered up and are difficult to locate.

5. Make connections to valve ports so that the connections do not leak, but do not apply so much effort in tightening the connections that the ports are broken.

6. Install valves so that they can be easily accessible to the operator, yet high enough above the floor that they cannot be bumped by lift trucks and carts.

7. When installing valves, do not use excessive pipe compound in the threaded ports. Be sure that no dirt has entered the valves before they are installed.

Mechanical Operation—When mechanically operated valves are installed these rules should be followed:

1. The valve should be mounted on a good flat surface. This valve must withstand thrust from cams, trip arms, pins, and other mechanical operating means. The mounting screws must be pulled down evenly and tightly.

2. If the valve is of the manifold type, be sure that the seal is in place.

3. Be certain that the valve is clean before it is installed.

4. Install the valve, so that it works best into the piping layout. In other words, eliminate bends in the piping as much as possible.

5. Make certain that all pipe connections are tight. Oil and air leaks are costly.

6. Check mechanical trips to see that they are in proper alignment. Also check for overtravel.

7. Lubricate the actuating mechanisms, and be sure that they operate freely.

Solenoid-operated Valves—When installing solenoid-operated valves, here are some of the points to follow:

1. Mount the solenoid valve per manufacturer's recommendations. Some designs are recommended for horizontal

movement of the plunger, and other designs are recommended for vertical movement of the plunger.

2. If the valve is of a type that is designed for pipe line mounting, make certain that the valve is piped upward, so that it does not leak. Since many in-line mounted valves have cast iron or brass bodies, make certain that the pipes are not screwed into the valve body to such an extent that they crack the valve body.

3. Check carefully the current specification on the nameplate of the valve or on the solenoid coil, before connecting the wires. Never use a different electrical current from that specified on the valve. The manufacturer's recommendations should be followed.

4. Keep solenoid valves away from regions where the temperature is high. Solenoid coils should be kept at normal temperatures if possible.

5. In explosive atmospheres, install solenoid valves which have explosion-proof housings that meet with Underwriter's approval for the conditions involved.

6. Protect solenoid valves, especially the type that do not have weatherproof housings, from water spray and excessive residue in the air.

7. Make certain that the interior of the valve is clean when it is installed.

8. Mount the solenoid valve so that it is protected from shop trucks, tote pans, and other items which may damage the valve.

9. Be sure that the solenoid covers are in place before the valve is placed in service; this keeps out dirt and foreign particles.

Pilot-Operated Valves—Some important points in installations of these valves are:

1. If the valve has mounting feet, mount it securely to a smooth flat surface. If the valve is pipe line mounted, the pipes should be securely tightened into the valve body.

2. Secure the pilot connections so that there is no leakage.

3. Make certain that the pilot connections are not restricted.

4. Keep the valve away from hot blasts.

5. See that lubrication is available to the valve.

6. Mount the valve, whenever possible, in a horizontal position. Although this is not necessary for some valves, it is necessary for others.

7. Do not hammer on the valve either when installing it or after it is installed.

8. If, when mounting the valve, the piping is not ready to install, do not remove the pipe ports.

9. Be sure to select the correct pilot valve for actuating the pilot-operated valve.

10. Do not connect pilot pressure into the spring end of a spring offset valve.

Causes Of Failure

Since there are so many different designs of valves, it is impossible to cover the causes of failure for each individual type of valve.

Manually Operated Valve—There are several things that may cause two-way valves to function improperly, such as:

1. *Dirt.* This is a major cause for valve failure. Dirt from unfiltered lines or dirty systems can score valve seals, scratch valve liners and metal pistons, clog the fluid passages, and do all kinds of damage. Dirt beneath a poppet causes leakage.

2. *Lack of lubrication.* Most two-way pneumatic valves require lubrication. Without lubrication, parts wear quickly, score, and shrink. A good light grade of spindle oil makes a good lubricant. Lack of lubrication often causes the actuator of a valve to stick.

3. *Heat.* This causes deterioration of most packings, which results in leakage. Heat can also cause a binding action in certain types of valves. This may be due to the swelling of packings and seals; or, in some valves, the binding action may be due to the rapid expansion of one part over another.

4. *Broken parts.* Some valves have springs for returning the actuator to the normal position. If the actuator does not

return, look for a broken spring. Broken diaphragms in a diaphragm type of valve cause a leakage problem. Other broken parts cause failure of the valve.

5. *Incorrect packings.* On a hydraulic valve, some fluids cause the packings to swell. Use packings that do not cause swelling with the fluid that is used.

Mechanically Operated Valve—If a mechanically operated valve fails to function, the points to check are:

1. If the valve fails to pass the rated amount of fluid, check to determine whether the actuating means is stroking the valve mechanism to the full extent of its stroke. A cam may have become loosened, a trip arm may have become bent, or the tripping mechanism may have become worn to such a point that it does not cause the valve to open fully.

2. If the toggle or cam arm is broken, check for overtravel on the tripping mechanism. This can cause considerable damage to the valve actuating mechanism. Pin-type actuators can also be damaged by overtravel on the tripping mechanism.

3. If the cam roller stem or a cam roller arm becomes bent check the rise on the cam. A cam with too steep a rise is bound to cause trouble. The more shallow rise causes less strain on the roller mechanisms.

4. If the cam roller, pin, or toggle (except the locking type) fails to return to the original position when pressure is released, several things should be checked. First, check for a broken return spring. Second, check for improper lubrication. If the valve has packings in it, some types of lubricants may cause the valve mechanism to stick, and the spring pressure may not be strong enough to return it to the starting position. Third, check for dirt; if the valve is located in air in an extremely dirty location, small particles of dirt may work downward into the operating mechanism and cause it to stick. Fourth, check for a binding action caused by side thrust from the mechanical operator,

which may have bent the valve mechanism and caused a binding action.

5. If the valve is a two- or three-way type and is connected to a pilot-operated valve, and when the two- or three-way valve is actuated and nothing happens, the fault may lie in the pilot-operated valve, rather than in the two- or three-way valve.

6. If a mechanically operated valve leaks excessively, dirt or heat, which may cause packing failures or excessive wear, may be the actual cause.

Solenoid Valves—Causes of failure of solenoid valves are:

1. If the valve mechanism fails to shift, a number of things may be wrong:

   a. The line pressure may be too high for the valve design. In a two-way valve, too much pressure may not allow the solenoid to raise the plunger to the valve seat. This is similar to lifting a weight. A 50-pound weight may be lifted without much effort, but if a 500-pound weight is attempted, it could not be lifted. Remember that valves are built for certain pressure ratings, and the manufacturer who built the valve knows that it will work satisfactorily at that pressure, with a margin of safety. Do not exceed his pressure recommendations.

   b. Many solenoid valve malfunctions can be traced to low voltage. Solenoid ratings are usually set up for 85 percent of full line voltage; but it is not uncommon for the voltage to drop under this point during peak loads, and malfunctions can be expected when this occurs. Check the voltage at the solenoid with a voltmeter.

   c. A binding action in the valve operating mechanism, due either to misalignment or to lack of lubrication, may be great enough to cause the solenoid to fail to shift.

   d. A burned-out coil will cause the valve to fail. This can usually be detected by the odor of a burning coil.

   e. A broken lead to the solenoid prevents current flow to the solenoid, and the solenoid will not operate.

    f.   Solenoid valve failure can often be attributed to a condition outside the solenoid, that is, a faulty limit switch, or an electrical actuator may not allow current flow to the solenoid, so that the solenoid can actuate the valve mechanism.

    g.   The use of incorrect voltage, that is, the voltage connected, may cause failure. The correct voltage for which the coil was built should be applied.

    h.   Broken parts, such as connecting pins or lever arms, can cause failure of the valve to shift.

    i.   If a solenoid cover is left off a solenoid, it may not function properly.

2. If the valve does not pass the expected amount of fluid, check to determine whether the solenoid plunger is opening the valve fully. Wear in linkage connecting pins and other parts may be the cause.

3. If the solenoid hums loudly and the coil burns out, the probable cause is the solenoid plunger not seating properly. This can be caused by dirt in the solenoid, a broken wearstrip, a loose rivet in the solenoid plunger, a condition in the valve actuating mechanism that causes incorrect voltage, or a condition in the valve itself.

4. If a solenoid hums, but does not cause trouble, loose laminations can be the cause.

5. If a solenoid valve with a spring-return feature fails to operate, check for a broken spring. Other things which may cause a failure to return the solenoid plunger are excessive fluid pressure, dirt, packing friction, and lack of lubrication.

6. Leakage in a solenoid valve may be attributed to many of the same things that occur in manual and mechanically operated valves, such as: dirt; scored seats; broken seals; lack of, or improper, lubrication; heat; and wear.

7. If the solenoid valve is cycled at too high a rate, the solenoid coil may be overheated, causing coil failure. Excessive shock and vibration due to high cycling of valves with large solenoids may cause mechanical failure in solenoid-operated valves.

Some solenoid valves are built with manual actuators, so that the valve can still be operated in event of power failure. A valve of this type is also advantageous in trying out a new circuit, where it is desirable to go through the cylinder before the electrical connections are made.

Pilot-Operated Valves—Causes of failure of these valves are:

1. In the main part of the valve, the causes of failure are similar to those that have been studied previously, such as dirt, heat, leaks, faulty packing, and broken springs.
2. If the seals in the pilot section are worn or if the cover gasket is leaking in a bleeder-operated pilot-operated valve, the valve mechanism will not shift properly. It has been found that if all the parts in the valve actuating mechanism are loose, the mechanism will shift when actuated, and then bounce back to the center position; then the valve becomes inoperative.
3. In a pressure pilot-operated valve, any leakage between the pilot valve and the pilot-operated valve may cause malfunction.
4. If dirt lodges in the pilot chamber, it can cause the pilot-operating mechanism to hang up, causing malfunction.
5. Lack of lubrication in the pilot chamber can cause scoring of the chamber wall, and cause the operating mechanism to stick and malfunction.
6. If the pilot operating mechanism is provided with a packing, an improper lubricant can cause the packing to swell, which causes the mechanism to seize.

Repair and Maintenance

Most two-way valves can be repaired. It is often less expensive to replace a plug-type valve with a new one. A manually operated two-way valve (see Fig. 3) is used here to illustrate repair procedure, as follows:

1. Use a clean workbench for disassembly of the valve. Place the valve in a vise, but do not exert too much pressure.
2. Use the following procedure for disassembly. If possible, secure a cross-sectional drawing of the valve to be repaired.

First remove the socket-head setscrew. Then loosen the valve cap with a wrench. Remove the valve cap which is connected to the valve body. The cup spacer, cup follower (upper), spring retainer, lock washer, button, return spring, piston rod, cup packing, washer, cup follower (lower), and piston nut can then be removed. Remove the valve body from the vise; place the button in the vise, and remove the piston nut. This completely disassembles all parts, except the lock washer, button, and piston rod. If these parts are not damaged, it is not necessary to disassemble them. Check all parts and clean them thoroughly. If the valve liner is worn or badly scored, it should be replaced by pressing out the old liner, and then shrinking in a new one. A burnishing tool should be used in the liner. Replace the cup packings with new ones.

3. Reassemble the valve, using lubricant on the parts, then test. Apply fluid pressure to the right-hand port, and check the left-hand port for leaks. If there is leakage, the cups are not seating properly; operate several times and recheck for leaks. If the leaks persist, it is necessary to disassemble the valve again and check the cup packing.

4. Check the stroke of the actuator to determine whether the required stroke can be obtained.

5. The valve is then ready to be placed back into the system. If the valve is to be placed in stock, be sure to seal the ports with a shipping plug to protect it from the elements and from dirt.

6. In assembling valves, always remember to use extreme caution to protect packing, seals, and gaskets from damage, as they can be a source of trouble.

7. Always keep moving parts well lubricated. Inspect valves at regular intervals for leakage and for wear in the actuator mechanism.

The repair of a mechanically operated two-way valve is similar to that of a manually operated valve. Also, it is nearly impossible to cover the repair of all types of solenoid valves. Except for the actuator, repair is similar to other two-way valves of like designs.

THREE-WAY VALVES

A three-way valve has three ports and three internal passages. One of the ports allows fluid to enter the valve; it is known as the inlet port. Another port allows the fluid to leave the valve and pass to some means of operation; this is usually known as the cylinder port. The third port allows the fluid to exhaust and is known as the exhaust port. In one of the positions of the actuator, the inlet passage is connected to the cylinder passage, and the exhaust passage is blocked off. In the other position of the actuator, the inlet passage is blocked, and the cylinder passage is connected to the exhaust passage.

Three-way valves, which have two inlet ports and one outlet port, are in use. In this type of valve, the fluid cannot exhaust backward through the valve. In one position of the actuator, the No. 1 inlet passage is connected to the output passage, and the No. 2 inlet passage is blocked. In the other position of the actuator, the No. 2 inlet port is connected to the outlet port, and the No. 1 inlet port is blocked. Different operating pressures may be used at each inlet port.

Three-way valves have many uses. Some of them are used to actuate single-acting cylinders, to actuate pilot-operated valves, and to bleed off pressure.

Actuation

The three-way valve is closely related to the two-way valve in basic construction. The same type of actuators are used, and, in many instances, the same valve mechanism is used for both valves.

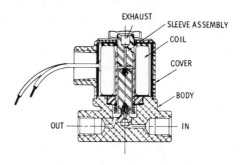

Fig. 23. Three-way solenoid valve.

The part names are much the same. The body in the three-way valve has one more port than that in the two-way valve. Three-way valves may be manually, mechanically, or pilot actuated. An electrically operated valve is shown in Fig. 23.

Maintenance of Three-Way Valves

The causes of failure in a three-way valve are much the same as in a two-way valve. In disassembly and maintenance of valves, a three-way valve of the poppet type (Fig. 24), can be studied

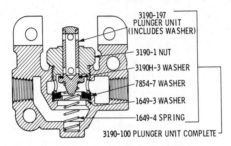

Courtesy Scovill Fluid Power Division, Scovill Mfg. Co.

Fig. 24. Poppet-type three-way valve.

for disassembly procedure. When the valve is disassembled, clean all parts thoroughly, and replace any worn parts. Examine the poppet seats for score marks. Replace poppet seals when necessary. Reassemble the valve, check it for leaks, and determine whether it functions properly.

FOUR-WAY VALVES

A four-way valve has four ports and four internal passages. As in the two-way and three-way valves, the number of ports determines whether the valve is a two-, three-, or four-way control valve. In the four-way valve, one port receives the fluid, and is known as the inlet port. Another port allows fluid to be exhausted, and is known as the exhaust port. Also, two cylinder ports, which usually direct the fluid to a cylinder, are known as cylinder ports. The cylinder ports are usually labeled No. 1 and No. 2. Four-way valves are used extensively in both pneumatic and hydraulic circuits. They are built in many port sizes, beginning at 1/8-inch to larger than 2 inches. The majority of valves in use are under 2 inches in size. Four-way valves for air usually do not exceed 150 psi in operating pressure, but hydraulic valves are often made for 5000 psi. The majority of hydraulic valves are used in the 1500 psi range. Four-way valves may be further classified as two-position and three-position valves.

A two-position four-way valve means that the actuator (or means of operation) has two positions. In one position, the inlet port is connected to cylinder port No. 1, and the exhaust is connected to cylinder port No. 2. In the other position of the actuator, the inlet port is connected to cylinder port No. 2, and the exhaust is connected to cylinder port No. 1.

A three-position four-way valve means that the actuator (or means of operation) has three positions—one of which is the center, or neutral, position. In the two extreme positions, the action is the same as in a two-position valve but in the neutral position, several combinations are possible. In one type of valve, all of the ports are blocked, which means that no pressure can get through the valve, and all passages are disconnected from each other, as shown in Fig. 25. In another type of action, when the actuator is in neutral position, the inlet is blocked, and both cylinder ports are connected to the exhaust port. It is conceivable that in still another type of action, the inlet and both cylinder ports could be exhausting. Other combinations are possible.

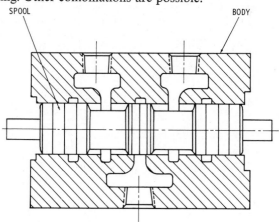

Fig. 25. Closed-center four-way control valve.

Four-way valves are used to actuate cylinders of the double-acting type and to actuate large four-way pilot valves. It is conceivable, and often done in an emergency, to block one port of a four-way valve, making a three-way valve—or to block two ports, making a two-way valve. Likewise, one port can be blocked in a three-way valve to make a two-way valve.

Actuators

The actuators for the two-position four-way valves resemble the two- and three-way valves, but the three-position four-way valve has a neutral position for the actuator. The manual actuator is shown in Fig. 26. This valve is suitable for 1500 psi oil operating pressure. The parts of this valve should be studied, since it typifies the basic design of many spool-type hydraulic four-way valves. Fig. 27 shows a four-way air valve of the double pilot bleeder type, which uses poppet assemblies. A hydraulic solenoid-controlled pilot-operated four-way valve with sub-plate mounting is shown in Fig. 28. Pilot-operated hydraulic valves of some designs may be actuated by air pressure.

A four-way pilot-operated control valve, depending on design, may be controlled by two-, three-, or four-way pilot valves. They are actuated as follows:

1. Four-way pilot-operated valve, pneumatic, of the bleeder type, in which the air is bled off the pilots to effect an actuation of the valve, is shown in Fig. 29. As the two-way

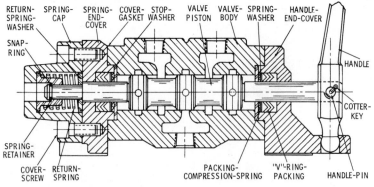

Courtesy Logansport Machine Co., Inc.

Fig. 26. Manually operated four-way hydraulic control valve.

valve (No. 1) is opened (see Fig. 29), the pressure is unbalanced in the four-way valve and the valve spool shifts toward valve No. 1, since there is pressure trapped between the four-way valve and valve No. 2. This pressure pushes over the valve spool. The opposing action takes place when valve No. 1 is closed and valve No. 2 is opened. This is often referred to as a bleeder type of pneumatic system.

2. Four-way pneumatic valves of the pressure type can be operated either by two three-way pilot valves or by one four-way pilot valve. Fig. 30 shows the use of two three-way valves.

To effect an operation with pressure to the pilot connection, a little more piping is required, depending on how

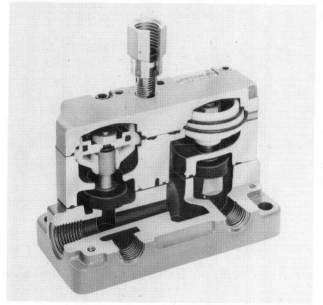

Courtesy The ARO Corporation

Fig. 27. Bleeder-operated four-way control valve.

far the pilot valves are located from the four-way valves. When the pilot valve is actuated, the air flows to the pilot connection, and the four-way valve spool is shifted. The spool moves very rapidly.

247

3. Three-position four-way hydraulic valves of the open-center type require either two- or three-way pilot valves; or a four-way pilot valve may be used to effect an operation. Fig. 31 shows how the four-way pilot valve controls the pilot-operated valve. Note the spring-loaded check valve, which is used to provide pilot pressure. The back pressure should be kept as low as possible in the hydraulic valves.

Courtesy Continental Hydraulics

Fig. 28. Four-way, solenoid-controlled, pilot-operated valve for sub-plate mounting.

Installation

Installation of four-way valves, in general, is similar to that of the two- and three-way valves. The valve should be solidly mounted in a location easily accessible to the operator.

When installing solenoid-operated four-way valves, check the electrical circuit, so that the solenoid is connected correctly to the switch or push button. It is well to check the circuit before the valve is piped to the operating cylinder. A hand can be held over each of the cylinder ports, and the solenoid tripped, to determine whether the fluid comes out of the correct ports.

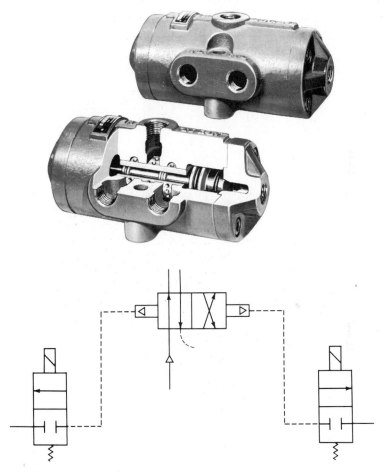

Fig. 29. Bleeder-operated four-way valve (top), as used in circuit (bottom).

When installing three- or four-way solenoid valves, make certain that the exhaust port is open, or connected to a piped ex-

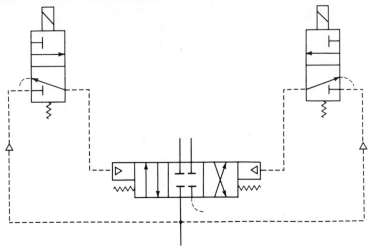

Fig. 30. Circuit including four-way pilot-operated valve controlled by two three-way valves.

haust that is unrestrained, except where a back pressure is required to operate the pilot valves.

Causes of failure for manually operated valves and mechanically operated valves, in general, are similar to that of two- and three-way valves. In a three- or four-way solenoid valve, too much fluid pressure, depending on valve design, places excessive pressure on packings and seals, and the force of a solenoid may not be enough to overcome this condition. Incorrectly installed directional control valves may prove to be expensive in terms of maintenance and repair costs.

Maintenance

A plug-type air control valve, (Fig. 32) is one of the more simple forms of a three-position four-way air valve. To disassemble a plug valve of the type shown, first remove the nut which holds the handle; then remove the handle. Next, remove the cover by unscrewing it, and the remainder of the parts can then be removed. Clean the parts thoroughly in a cleaning solution. Check the tapered seat in the body for score marks. If the scores are not too deep, they can be polished out. The plug should then be

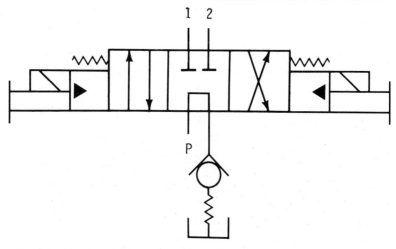

1 2

P

Fig. 31. *Circuit including a spring-centered pilot-operated control valve.*

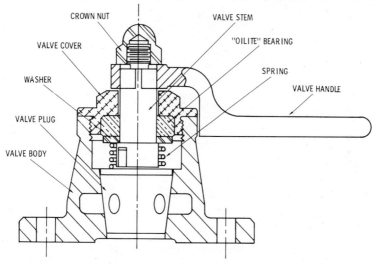

CROWN NUT

VALVE STEM

VALVE COVER

"OILITE" BEARING

SPRING

WASHER

VALVE HANDLE

VALVE PLUG

VALVE BODY

Courtesy Logansport Machine Co., Inc.

Fig. 32. *Plug-type air control valve.*

checked and polished, if it is scratched. The plug can be lapped
into the body with a lapping compound, but be sure that all the
compound is cleaned off before the valve is reassembled. Grease
the plug and seat it in the body. Grease the bearings and reas-

251

semble the valve, making certain that all parts are properly re-placed. Check for leaks by shifting the handle to the neutral position and applying fluid pressure to the inlet. Place a hand over the other three ports. Next, plug cylinder port No. 1, shift the fluid pressure to the seat port, and check the exhaust and cylinder port No. 2 for leaks. Switch the handle to neutral, remove the plug from port No. 1, and plug port No. 2. Shift the fluid to port No. 2, and check port No. 1 and the exhaust ports for leaks. Also, check the bottom of the plug and around the cover to determine whether any fluid is escaping. If there is an excessive amount of leakage, disassemble and further lap the plug into the body.

If the valve (Fig. 33) is to be disassembled while it is in the machine, the air pressure and the electric current should be shut

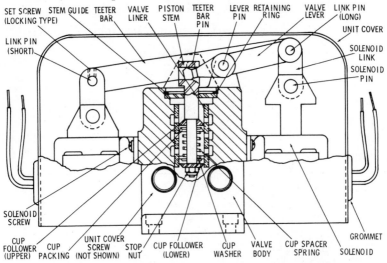

Fig. 33. Solenoid-operated four-way air control valve.

off. To remove the valve cover, remove the screws and pull off the cover, being careful to avoid damage to the solenoid lead wires. Wash the cover and check the grommets to determine whether they need replacing. Loosen the setscrews, and then re-move the link pin (short), teeter bar pin, lever pin, solenoid pin, link pin (long), solenoid link, valve lever, and teeter bar. Remove the solenoids by removing the solenoid screws. To disassemble the valve, remove the retainer ring from the valve body and remove the piston assembly which is composed of the piston stem, cup follower (upper), cup packing, cup spacer spring, cup follower (lower), cup washer, and stop nut; then the stem guide can be pulled out of the valve body. To remove the cup washers, place the piston assembly in a vise with the top of the piston assembly placed firmly in the vise jaws. Remove the stop nut; then all the parts of the piston assembly can be removed. Clean all parts thoroughly with a solvent to remove dirt and hardened grease. Inspect each part for wear or damage. Place all the parts in a clean container to prevent losing any of the parts.

If the valve liner is scored, it should be pressed outward and replaced. A new liner should be pressed in to the correct depth; then a burnishing ball should be pressed through the liner, which is 0.002 to 0.003 inch over actual bore size. Check the solenoids with electric current to determine whether they are damaged. If there are damaged parts in the solenoids, they should be replaced with new parts.

In reassembling the piston, place the following parts in order on the piston stem: stem guide, cup follower (upper), cup packing, cup washer, cup spacer spring, cup washer, cup packing, cup follower (lower), and stop nut. Make certain that the stop nut

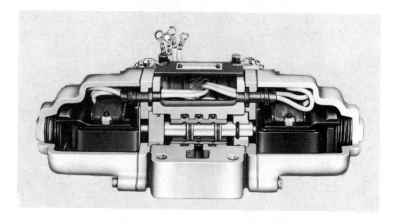

Courtesy Logansport Machine Co., Inc.

Fig. 34. Subplate-mounted solenoid-operated air valve.

is tightened securely and that the lips of the cup packing face each other. Place a light cup grease on the cup packings, and place the piston assembly into the valve. Lock the piston assembly into place with the retainer ring. Hook the intake of the valve to the air line, and check for leaks by plugging both cylinder ports, placing one hand on the exhaust port when the valve stem is in the extreme positions. If there is no exhaust leakage, continue to assemble the entire unit, reversing the sequence of disassembly. When replacing the solenoid, make sure that the solenoid screws are tight. Lock all setscrews securely, and grease all pins with a lubrication that will not drip off. Before replacing the cover, determine whether the valve operates freely by manual means. Then hook up the air supply, keeping both cylinder ports blocked. Next, hook up the electric power, and check the operation of the valve. If the valve works satisfactorily, disconnect the air and electricity. Push the electric leads through the grommets

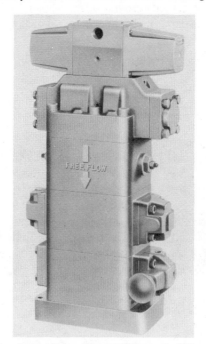

Courtesy Sperry Vickers, Division of Sperry Rand Corporation

Fig. 35. This control package saves space in hydraulic systems.

in the cover, and fasten the cover with the cover screws. If the valve is to be placed in storage, plug the ports and store in a cool dry place. If the valve is to be placed in service, be sure that it is sufficiently lubricated. Check the valve regularly for leakage and performance. If the valve is subject to hot blasts, protect it with a shield or baffle.

Many valves, especially solenoid-operated valves are now designed for subplate or manifold mounting. Some solenoid valves also have plug-in connections for the solenoids. These features make them easier to install and to remove, as the piping is not disturbed. Fig. 34 shows an air valve that has both the subplate mounting and the plug-in solenoid connections. Note that the spool of the valve is operated directly by the solenoids. Non-metallic inserts are installed in each end of the spool, so that the plunger pins on the solenoids do not mushroom the ends of the spool. The covers on the valve contain the solenoids, so that if the covers are removed, the solenoids drop off and make the valve inoperative.

A compact hydraulic control installation is shown in Fig. 35. At the top is a four-way control valve with a solenoid-operated pilot valve. Directly underneath is a flow control, and a check valve is underneath the flow control. Adjacent to the subplate is a relief

Courtesy Commercial Shearing, Inc.

Fig. 36. This four-way control has built-in relief valves and a check valve.

valve. This installation eliminates considerable high-pressure hydraulic piping and saves space.

A cross section of a hydraulic four-way control valve is shown in Fig. 36, with an adjustable cross-over relief valve in each of the two cylinder ports and a check valve in the parallel passage. The cross-over relief valve protects that portion of the circuit piped to any working section against overload. In this setup, high-pressure oil is allowed to cross over from one work port to the other work port. These relief valves function when the spool of the working portion is in neutral position, since the setting of cross-over relief valves is generally higher than the main relief valve, and the cross-over relief valves are not functional when the spool of the working section is in the work position. This concept of the relief valves and the check valve built into the main body of the valve eliminates considerable piping and permits a cleaner and more efficient installation.

REVIEW QUESTIONS

1. What is meant by a three-way valve?

2. What is the chief difference between a three-way valve and a four-way valve?

3. What is meant by a directional control valve?

4. What is meant by a two-position four-way valve?

5. What is the chief difference between manual and mechanical operation of a valve?

6. Explain the operation of a pilot-operated valve.

7. What is meant by a solenoid valve?

8. What factors may cause the failure of a valve?

9. What size (pipe size) of valves should be used to pass the following volumes of oil: 3 gpm, 8 gpm, 20 gpm, 35 gpm, 75 gpm, 100 gpm, 150 gpm, 250 gpm?

10. Draw the *ANS* Graphic Symbol for each of the following pneumatic valves:

(a) Solenoid-operated, normally-open two-way valve.

(b) Manually operated three-way valve.

(c) Three-position, spring-centered, closed-center, solenoid-operated four-way valve.

(d) Air-piloted, two-position four-way valve.

11. Draw the *ANS* Graphic Symbol for each of the following hydraulic control valves:

(a) Two-position, pilot-operated, solenoid-operated four-way valve.

(b) Three-position, closed-center, manually operated four-way valve.

CHAPTER 12

Flow Controls

The function of a flow control is to control the volume of fluid that passes a certain point in the circuit. A flow control is actually a valve. A flow control might be compared to the heat register in a home. If it is opened only a small amount, only a small quantity of hot air can come out; but if it is opened widely, the hot air rushes out. The register can be regulated from completely closed to fully open in small steps.

A flow control is used to control the piston speed of power cylinders, the speed at which the spool of a pilot valve shifts, the timing cycle on a timing valve, and the speed at which the shaft of a fluid motor revolves. A flow control is a simple device, but it plays a very important role in a fluid power circuit.

Since both air and hydraulic flow controls are studied in this chapter, the types of flow control valves that are used in industry should be studied also. Although many pneumatic flow controls are designed for air pressures up to 150 psi, they are also used for low-pressure oil service up to 500 psi; it is recommended that pneumatic flow controls be used for pneumatic service and that hydraulic flow controls be used for hydraulic service.

TYPES OF FLOW CONTROLS

Flow controls can be divided into three types:

1. Metering-in type (Fig. 1).
2. Metering-out type (Fig. 2, Fig. 3, Fig. 4).
3. Metering-in-both-directions type (Fig. 5).
4. "Bleed-off" type (Fig. 1).

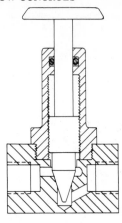

Fig. 1. Metering-in type of valve.

In the metering-in type of flow control, the fluid is metered on its way into the device that it is going to control (see Fig. 1). The metering-out type of flow control meters the fluid as it leaves the device that it is going to control (see Fig. 2). This is often called exhaust metering. The metering-in-both-directions type of flow control meters the fluid in both directions (see Fig. 5).

Most of the meter-in type of flow control valves are used for hydraulic service, and many of the meter-out type of flow control valves are used for pneumatic service. Many of this type are also used for hydraulic service. The meter-in-both-directions type of flow control valves is not used to a great extent in industrial fluid power systems. "Bleed-off" types of control valves are used in hydraulic systems; they are used to bleed off a predetermined amount of oil, which is under pressure.

Flow controls are also built in conjunction with cam-operated shutoff valves which allow the fluid to flow freely until the cam actuates the shutoff valve; then the fluid must flow through the flow control. Flow controls are built with adjustable orifices (Fig. 6), that is, the passage through which the fluid passes can be made either larger or smaller to take care of various requirements. The advantage of this arrangement is that the movement of the device which the flow control is controlling can be either speeded up or slowed down. Even in the same application when different loads are applied, it is often desirable to be able to change

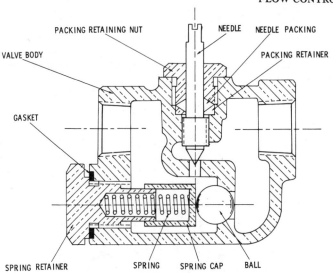

PACKING RETAINING NUT — NEEDLE — NEEDLE PACKING

VALVE BODY — PACKING RETAINER

GASKET

SPRING RETAINER — SPRING — SPRING CAP — BALL

Fig. 2. Illustration of a pneumatic flow control.

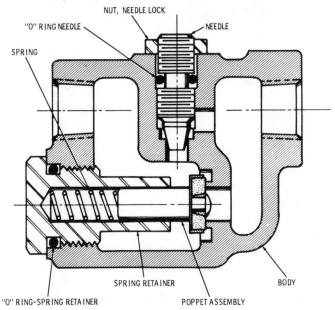

NUT, NEEDLE LOCK

"O" RING NEEDLE — NEEDLE

SPRING

SPRING RETAINER — BODY

"O" RING-SPRING RETAINER — POPPET ASSEMBLY

Courtesy Logansport Machine Co., Inc.

Fig. 3. Pneumatic flow control (also used for low-pressure oil).

261

the flow for each load. For example, the flow control may be used to regulate the speed at which a saw is being fed into lumber. When kiln-dried lumber is being sawed and a change is made so that green lumber is being sawed, it is usually desirable to change the speed at which the saw is fed into the work.

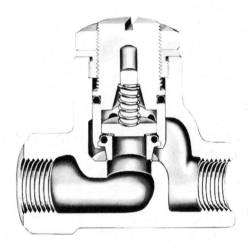

Courtesy Scovill Fluid Power Division, Scovill Mfg. Co.

Fig. 4. Cutaway view of a flow control for air, oil, or water service.

Flow controls which either meter-in or meter-out usually have another valve built in conjunction with them, that is, a check valve. This valve allows for free flow when the fluid is flowing in the opposite direction.

The cam-operated valve, (Fig. 7) is really three valves built into a single valve—a shutoff, a flow control, and a check valve. This type of valve offers a distinct advantage, because the means which the flow control is regulating can be fed rapidly to the work; then when the shutoff valve is closed, the flow control takes over for the work cycle. A typical example is the tool feed on a machine tool. The cylinder moves the piston and the tool inward rapidly until the cam roller is depressed; then the piston and tool

Fig. 5. Flow control for metering in both directions.

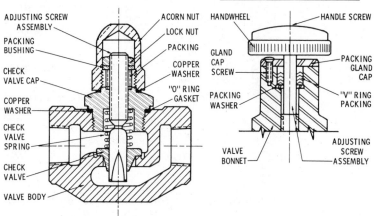

Courtesy Logansport Machine Co., Inc.

Fig. 6. Hydraulic flow control valve.

are moved at the rate set by the flow control. If a skip feed is desired, short cams can be used on the feed table (Fig. 8).

In hydraulics, panel valves which contain flow controls and various operating valves are used in complicated machine feed circuits. The cam-operated flow control valve can be used advantageously where a long cushion is desired on a cylinder.

The check valve in the flow control valves is a device that allows the fluid to flow freely in one direction, but not in the other direction. This action is similar to that of a door in a home; it cannot be pushed open in one direction because of the door jamb, but it can be pushed wide open in the other direction.

Pneumatic flow controls are usually available in sizes from 1/4-inch to 1-inch (pipe size), and hydraulic flow controls are avail-

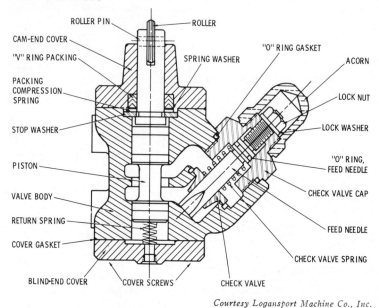

CAM-END COVER

ROLLER PIN

ROLLER

"V" RING PACKING

SPRING WASHER

"O" RING GASKET

ACORN

PACKING
COMPRESSION
SPRING

LOCK NUT

STOP WASHER

LOCK WASHER

PISTON

"O" RING,
FEED NEEDLE

VALVE BODY

CHECK VALVE CAP

RETURN SPRING

FEED NEEDLE

COVER GASKET

CHECK VALVE SPRING

BLIND-END COVER

COVER SCREWS

CHECK VALVE

Courtesy Logansport Machine Co., Inc.

Fig. 7. Cam-operated flow control valve.

able in sizes from 1/4-inch to 2-inch (pipe size), and larger. Pressure ranges up to 150 psi are available for air, and they are available up to 3000 psi for hydraulics.

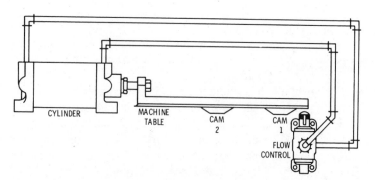

CYLINDER

MACHINE
TABLE

CAM
2

CAM
1

FLOW
CONTROL

Fig. 8. Use of cams on the machine table for operating a cam-operated flow control valve.

PART NAMES

The names of the major parts in flow control valves should be learned. Cross sections of typical flow controls are shown in Fig. 3 and Fig. 4. A pneumatic flow control is illustrated in Fig. 2, and a hydraulic flow control is diagrammed in Fig. 6. The following parts should be studied:

1. *Valve body.* All flow control valves have some type of body or housing. For pneumatics, the housing is usually made of a nonferrous material, so that condensation in the air line does not affect it. Brass, bronze, and aluminum are most often used. For hydraulics, the body may be made of high-tensile cast iron, cast steel, or steel bar stock. Although the finish on the outside of the body is not important, the finish on the inside is very important. The reason for this fact is that the seats and recesses for seals on the inside of the body must be leakproof. The body houses all valve parts, and contains the pipe ports.

2. *Orifice control.* The orifice control may be in the form of a needle, or another device which gradually closes or opens the orifice in the valve. This device is usually made of hardened steel. On the end opposite the closing device, a flat, a screw-drive slot, or a socket for a wrench is placed to facilitate adjustment. To open the orifice, turn the orifice adjustment in the counterclockwise direction.

3. *Orifice control stem packing.* This packing keeps the fluid from escaping between the stem and the valve body. The fluid is often under pressure. A hydraulic valve leakage at this point creates a "messy" condition. The stem packing is made of various materials, and is in various shapes. The packing materials used are rubber, synthetic materials, asbestos impregnated with graphite, and other materials. Shapes are in the form of an "O" ring, "quad" ring, Vee rings, formed wedge, and rope which can be formed as the retainer is seated.

4. *Check.* In order to obtain free flow in one direction in a flow control valve, it is necessary to have a built-in check valve in the valve body. The check valve may be in the

form of a plunger, a poppet, or a ball. In Fig. 2, a ball-type check valve is shown; and in Fig. 3, Fig. 4, and Fig. 6, a poppet-type check valve is shown. The check must fit the seat without any leakage, otherwise the check is not effective. The check in a pneumatic valve must be able to resist corrosion.

5. *Spring.* The spring is used in conjunction with the check to keep the check seated when fluid pressure is not attempting to open it. The spring helps the check to snap shut when pressure is released from the side opposite the spring.

Depending on the design, other parts of flow control valves are: lock nuts—used to lock the orifice control stem; cover for orifice stem, which can be fastened with a sealed lock wire so that the orifice setting cannot be changed unless the lock wire is broken and the orifice stem cover removed; gaskets, which seal the joints in the valve; and other small parts.

In a cam-operated flow control (Fig. 7), other parts are included, such as:

1. *Cam roller.* This is a hardened steel roller which is contacted by some mechanical means, such as a cam. The cam roller is fastened to the valve spool by a pin.
2. *Cover.* The cover supports the valve stem, and provides a guide for it. It is fitted closely to the valve stem.
3. *Valve stem.* This is part of the valve spool which, in a cam-operated valve, causes the normal flow passages to be closed off when the valve stem is depressed.

INSTALLATION

When installing this component in a fluid power system, certain items should be checked.

1. Check to determine whether the control is a metering-in or a metering-out type of flow control.
2. Check the circuit for the direction in which the motion is to be controlled.

3. Remove pipe plugs or pipe closures. Make sure that all of the pipe closure is removed.
4. Check to determine whether the pipe threads in the valve are clean.
5. Screw the pipe connections securely into the valve. High-pressure connections must be tight.
6. If a cam-operated valve is being installed, place some lubricant on the roller and roller pin.
7. When the fluid power is turned on, check the system thoroughly for leaks.

CAUSES OF FLOW CONTROL VALVE FAILURE

There are several causes of flow control failure as follows:

1. *Dirt.* Dirt which is trying to pass through the valve causes considerable damage to either the pneumatic or the hydraulic flow control valve. Dirt can clog the orifice; the result is either a stoppage of flow or a very erratic feed, depending on how much dirt is trying to move through the orifice.

   Dirt can cause leakage of the check by lodging on the check seat and causing the check to remain open when it should be closed. This condition also causes erratic feed. Dirt also can score the check valve seat, or orifice control, resulting in inaccurate control.

2. *Broken spring in check valve.* This often causes the check to seat improperly, causing an internal leak. It also can cause erratic feed, because the trapped air escapes through the check valve.

3. *Valve mounted in circuit backward.* Although this may not cause the valve itself to fail, it may cause the circuit to function improperly, causing circuit failure. This condition can be detected if no control is found in the desired direction. Most valves have an arrow which marks the direction of controlled or free flow.

4. *Improper valve for the installation.* Sometimes a low-pressure air valve is used for high-pressure hydraulic applications; the result is improper control, blown packings, and

blown gaskets. Be sure to select the correct valve for the application. An air valve should be used for an air application, and a hydraulic valve should be used for a hydraulic application.

5. The use of an *improper lubricant or oil* may cause trouble. In an air valve, certain lubricants may cause packing failure, and some types of hydraulic oils may cause packing failure.

6. If the *drain plug is not removed* in the hydraulic cam-operated flow control valve, the oil fills the lower part of the valve; then the cam roller cannot be depressed.

7. If the *oil in a hydraulic sytem becomes too hot,* gasket and packing failure may result. Touching the valve may determine whether it is overheating. Be careful when touching the valve, as it may burn a hand.

REPAIR AND MAINTENANCE

Dismantling of the single-unit flow controls is rather simple, as shown by the two valves in Figs. 2 and 4. The valve in Fig. 2 should be placed in a vise, so that both of the hexagon plugs are accessible. Remove the spring retainer with a wrench, being careful not to let the spring, spring cap, or ball fall to the floor. Check the ball carefully for dents or deep scratches. Check the tension on the spring. If these parts are undamaged, place them in a clean box. If any of these parts are dirty, wash them in a solvent before placing them in a box. Then remove the packing retainer nut, packing, and packing retainer. Next, remove the needle with a wrench or screwdriver. Wash the body thoroughly in a solvent. Check the needle seat and the ball seat for scratches or marks. If the ball seat is damaged, a ball can be dropped onto the seat and tapped until a new seat is formed. The needle seat can be polished, and the needle screwed in tightly to form a new seat. When reassembling the valve, replace the gaskets and packings to prevent leakage.

When the needle is screwed down, the air should be closed off. Check for leaks around the needle stem and the spring retainer. If there is leakage around the stem, tighten down on the packing. If leaks occur around the spring retainer, tighten it. Force air into

one port, and check to determine whether the air flows freely
from the second port.

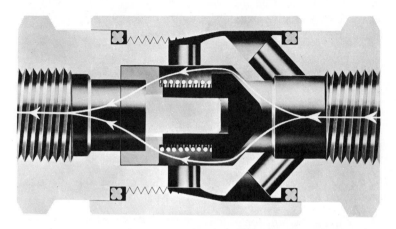

Courtesy Rexnord, Inc., Air Valves Division

Fig. 9. Flow control for either pneumatic or hydraulic service.

In valves containing synthetic seals (Fig. 9), use care in working the seals over the threads, so that the seals are not cut. It
may be necessary to wax threads that are extra sharp to prevent
cutting the seals. After the valve is reassembled, it should be set up
and tested—force air into one port, and open and close the needle
to determine whether it works properly.

If the flow controls are to be placed in storage, be sure that the
ports are plugged. Store in a clean, dry, and cool location. Tag
the valve to show date of repair. In operation, flow controls require very little maintenance, but they should be checked occasionally for leaks.

Fig. 10. Quick-exhaust valve.

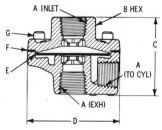

Courtesy Scovill Fluid Power Division, Scovill Mfg. Co.

Fig. 10 shows a quick-exhaust valve which may be considered a flow control, since it dumps the exhaust air into the atmosphere and allows much faster movement of the piston in a cylinder. By using this type of valve, piston movements may be increased threefold.

REVIEW QUESTIONS

1. What is meant by a flow control?

2. Give examples of flow controls that may be found in a home.

3. Describe a metering-in type of flow control.

4. Describe a metering-out type of flow control.

5. What is the function of a check valve?

6. What is the function of the spring in a flow control valve?

7. What things should be considered when installing a flow control?

8. What are the possible causes of failure of a flow control valve?

9. Give examples of flow control that may be found on an automobile.

10. Describe the installation procedure for a cam-operated hydraulic flow control valve.

11. How can a cam-operated flow control be used to advantage in a hydraulic system?

CHAPTER 13

Pressure Controls

The function of pressure controls is to control pressures in the fluid power system. Pressure controls may be used to reduce, to relieve, or to adjust pressure, or to begin another function.

Types of components that are considered to be pressure controls are:

1. *Relief valves.* In hydraulics, the function of the pressure relief valve is to relieve the pump to protect it and the system from becoming overloaded. When the pressure reaches a certain point, the relief valve spills the oil back to the reservoir, and the pressure is relieved. There are many modifications of these valves, depending on the system in which they are used.

   In pneumatics, the relief valve protects the system from exorbitant pressure, and bleeds off the pressure; it acts as a pop-off valve. A relief valve may be found on the hot-water tank in a home. It acts as a safety device in the home. The relief valve on the air or hydraulic system serves a similar purpose.

2. *Pressure reducing valve.* The pressure reducing valve is used to reduce the pressure in part of the system to a lower pressure. In a hydraulic system, it is often desirable to reduce pressure in certain parts of the system.

   The pressure regulating valve, as it is referred to in a pneumatic system, takes care of the pressure in the entire system. It smooths out the surges, and can be regulated in small amounts to give whatever reduced pressure is desired. This valve was discussed in Chapter 10.

3. *Sequence valve.* The function of a sequence valve is to set

up the sequence of operations in either a pneumatic or a hydraulic circuit. Sequence valves are also used for other functions. This is explained later in this chapter.

4. *Unloading valve.* The function of an unloading valve, as used in hydraulic circuits, is to unload the pressure at the desired instant, in order to conserve horsepower and to afford protection to the system. This helps to reduce heat.

THE HYDRAULIC PRESSURE RELIEF VALVE

The hydraulic pressure relief valve may be of the direct-acting type, the direct-operated pilot-type, or the remote-actuated pilot-type of valve. In the direct-acting valve, the fluid pressure acting on the piston must overcome the tension applied by a large spring in order to open the exhaust port. The direct-operated pilot-type is pilot-operated and uses only a small spring. The remote-actuated pilot-type is controlled by a remote valve through a pilot connection. It protects the pump, electric motor, fluid lines, directional controls, other controls, cylinders, and fluid motors from being overloaded, or from having an operating pressure applied that is above the safe range of the components.

Although hydraulic relief valves are usually considered to be part of the power device (Fig. 1), they may be used in other places in the hydraulic system, such as to relieve pressure during a stand-by period. A stand-by period is referred to as the idle time when a pneumatic or hydraulic press machine is not doing work, such as during the loading period. The advantages of using an extra relief valve in the circuit to take care of the stand-by period are that it reduces heat, reduces the power consumption, and takes the full load off the system for fairly long intervals. Relief valves are known as a normally closed valve, because the exhaust passage is kept closed until the piston opens that passage to relieve fluid pressure.

Hydraulic pressure relief valves of the direct-acting type are usually built for pressures up to 3000 psi; however, in some instances, they are designed for much higher hydraulic pressure. Usually, they do not cover the full range, but are found in ranges, such as: 50 to 750 psi, 700 to 1500 psi, 1500 to 2500 psi, and 2000 to 3000 psi. To understand the principle on which the

Fig. 1. The relief valve is a component of a power unit.

direct-acting relief valve functions, Fig. 2 should be studied. The fluid flows unobstructed in one port and out the other port, until a resistance is met; then the pressure inside the valve builds up to a point where the differential area between the top section and the bottom section of the valve piston, multiplied by the internal pressure in the valve, works against the spring. The piston rises to a point that allows the fluid to escape to the third port, relieving the pressure. The pressure against the piston depends on the tension placed against the spring by means of the adjusting screw.

This type of valve is built in port sizes ranging from 1/4 inch to 2 inches, or more. The part names, as shown in Fig. 2, should be studied. The piston is fitted closely to the valve body in order to reduce leakage to a minimum. The valve is simple in construction, having only two moving parts—the piston and the spring. The piston moves very rapidly.

The internal or external pilot-operated type of relief valve may be built for pressure up to 5000 psi, in sizes that are similar to the direct-acting type.

273

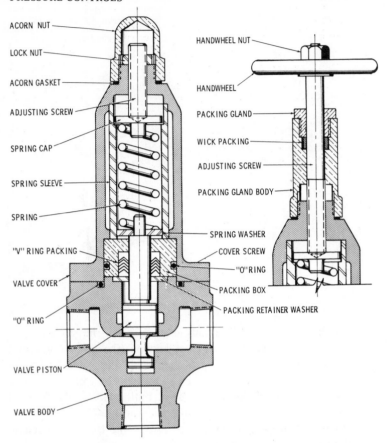

ACORN NUT

LOCK NUT

ACORN GASKET

ADJUSTING SCREW

SPRING CAP

SPRING SLEEVE

SPRING

"V" RING PACKING

VALVE COVER

"O" RING

VALVE PISTON

VALVE BODY

HANDWHEEL NUT

HANDWHEEL

PACKING GLAND

WICK PACKING

ADJUSTING SCREW

PACKING GLAND BODY

SPRING WASHER

COVER SCREW

"O" RING

PACKING BOX

PACKING RETAINER WASHER

Courtesy Logansport Machine Co., Inc.

Fig. 2. Direct-acting relief valve.

The direct-operated hydraulic relief valve is compact, since it does not require space for a large spring. Note that in the "M" series (Fig. 3) the spring is relatively small. The movable main poppet allows a large volume of oil to escape to the reservoir when the system pressure of the valve is reached. The action of the large main poppet is controlled by a much smaller poppet. The system pressure acts on both sides of the main poppet because of the small orifice. Since a greater area is exposed to system pressure on the top or left-hand side, the main poppet is held firmly on its seat, thus reducing leakage. System pressure also acts on the control poppet by way of the orifice just mentioned. When pressure becomes great enough to overcome the adjustable spring pressure bearing on the control poppet, fluid flows to the

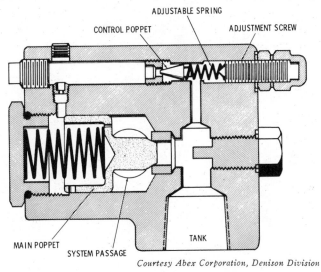

Fig. 3. Direct-operated pilot-type relief valve.

Courtesy Abex Corporation, Denison Division

reservoir. Forces are then upset on the main poppet, because flow past the control poppet causes a pressure difference across the orifice. Pressure on the bottom or right-hand side then exceeds that on the top or left-hand side and the main poppet moves upward, or to the left, off its seat. A large volume of oil can then escape to the reservoir at atmospheric pressure, thus reducing pressure in the system. When reduced pressure allows the control poppet to reseat, the main poppet again closes. The adjustment of pressure is made by means of a socket-head screw. In the valve, one spring is used for the entire pressure range up to 2000 psi. Some direct-operated pilot-type valves use the spring in increments, similar to the direct-acting type of hydraulic pressure relief valve.

The parts which make up the valve, as shown in Fig. 3, can be studied in the parts list shown in Fig. 4. The pilot-operated "RV" series relief valve, as shown in Fig. 5, is rated for 5000 psi.

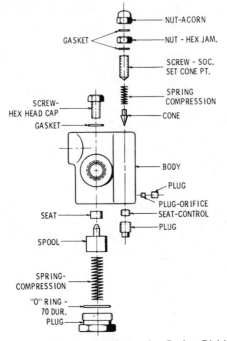

NUT-ACORN

GASKET

NUT - HEX JAM.

SCREW - SOC. SET CONE PT.

SCREW- HEX HEAD CAP

SPRING COMPRESSION

GASKET

CONE

BODY

PLUG

PLUG-ORIFICE

SEAT

SEAT-CONTROL

PLUG

SPOOL

SPRING- COMPRESSION

"O" RING - 70 DUR.

PLUG

Courtesy Abex Corporation, Denison Division

Fig. 4. Names of parts of the relief valve shown in Fig. 3.

Large-capacity relief valves are often built with flange-type connections. These valves are generally used where large volumes are concerned. Relief valves are also built for gasket mounting.

Installation

Relief valves that are not part of a power unit should be installed where they are readily accessible for servicing and for adjustment. It may be necessary for the setup man to make the adjustments.

Since many relief valves are pipe line mounted, make certain that they are connected firmly to the piping, so that no leakage occurs around the pipe threads. If they are gasket mounted, make sure that the gaskets are in place before tightening the mounting screws.

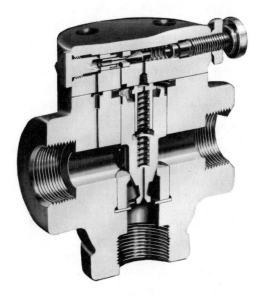

Courtesy Abex Corporation, Denison Division

**Fig. 5. Cutaway view of direct-operated pilot-type
relief valve rated for 5000 psi.**

Always install a return line to the reservoir. Make certain that
this line is of sufficient capacity and without undue restrictions in
order not to create back pressure.

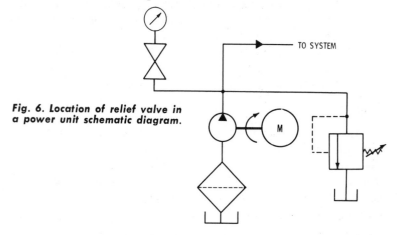

**Fig. 6. Location of relief valve in
a power unit schematic diagram.**

TO SYSTEM

277

Install relief valves that are of suitable capacity and strength to handle the pump in the system. Install the relief valve so that it is above the oil level in the reservoir. This allows the oil to drain out of the valve readily. Fig. 6 shows where the relief valve is placed in the power-unit diagram. After a relief valve is installed and set, the adjustment should be locked to keep anyone, except authorized personnel, from making changes in the pressure setting.

Relief Valve Failure

A number of conditions may cause a relief valve to fail to function properly:

1. *Broken valve spring.* This is usually detected by inability to change the operating pressure by changing the spring tension. This condition may also be caused by an open valve in the system.
2. Dirt often causes the piston in the valve to stick. Dirt also clogs small orifices, stopping oil flow through certain passages. Dirt may cause a fluctuation which can be noticed on the pressure gauge.
3. Air in the system also fouls the relief valve, causing pressure fluctuation and extreme noise in the relief valve.
4. The pump pressure and volume may be too high for the valve. Use good judgment in selecting the valve.
5. Internal leakage may occur when the valve piston and body become worn to a certain degree; the fluid passes from the inlet port to the exhaust port without causing the piston to move. In this instance, the valve should be replaced.
6. Back pressure often causes a relief valve to malfunction.
7. Excess heat may cause excessive leakage, or may cause the piston to stick.

Maintenance

After disassembling the relief valve, make certain that it is properly reassembled. Use caution in replacing cover gaskets, so that the gasket does not block off any passages. When replacing broken springs, be sure to use one that is recommended by the manufacturer of the valve. If the valve piston is loose in the body, replace the piston with an oversized one, or chromeplate the old piston so that it fits the body. Replace any worn-out packing before reassembling the valve.

Test the valve after it is rebuilt by connecting the inlet port to the pump, and plugging the outlet port. Connect the exhaust port to the reservoir, then change the setting of the adjusting screw to determine whether the full pressure range can be obtained in the system.

Since relief valves are very simple in construction, they usually give long troublefree service, unless dirt, air bubbles, and sudden shocks occur in the system. If these occur, much trouble can result.

PRESSURE REDUCING VALVE

The hydraulic pressure reducing valve may be either the direct-acting type or the direct-operated pilot-type. The direct-acting type is shown in Fig. 7. The direct-operated pilot-type is somewhat more compact, since it does not use the large spring (Fig. 8).

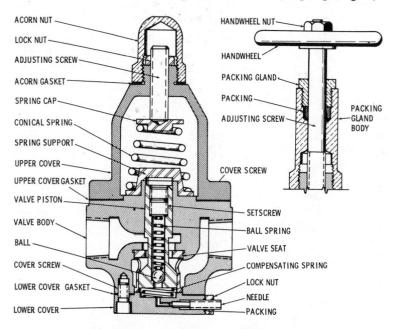

Fig. 7. Direct-acting type of reducing valve.

279

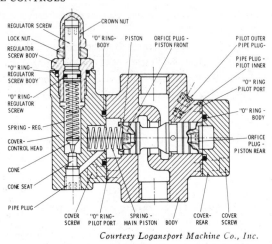

Courtesy Logansport Machine Co., Inc.

Fig. 8. Direct-operated pilot-type reducing valve.

In many hydraulic circuits or systems, more than one operating pressure is desirable in the system. The reducing valve may be the solution. In the hydraulic reducing valve shown in Fig. 7, when the spring tension is relieved from the large spring, the pressure differential is highest between the inlet port and the outlet port. As the tension is increased on the spring, the difference is reduced. The pressure differential may be as high as 10 to 1. For example, the inlet pressure may be 500 psi, and the outlet pressure may be 50 psi.

Reducing valves are built in pressure ranges similar to the pressure ranges of relief valves. They are usually not built in as many different sizes.

The parts which make up a hydraulic reducing valve (see Figs. 7 and 8) should be studied. Note in Fig. 7 that a check valve is built in, so that the oil can flow freely in the reverse direction, which eliminates the need for a separate check valve. The piston is hardened, and is lapped into the valve body. The valve seat is also hardened. The valve in Fig. 8 does not have the built-in check valve.

Installation

Pressure reducing valves should be installed where they are accessible for servicing and for making adjustments by the setup

280

man. Where valves are designed with locking wires, these valves should be sealed after the system is checked and released for production operation. This discourages operators from tinkering with the adjustment.

Since these valves are often pipe line mounted, they should be mounted securely in the pipe line. These valves, as well as all other valves that have pipe ports in their bodies, should be piped so that they are easily removed from the system. Reducing valves are also built for subplate mounting.

Connect the drain port to the exhaust lines. Make certain that there are no obstructions in the exhaust line.

Use a reducing valve with the correct pressure range for the system involved. Mount the valve so that the drain line is not subject to back pressure.

Causes Of Failure

Many of the same conditions that cause a hydraulic relief valve to malfunction also cause a hydraulic reducing valve to malfunction—dirt, internal leakage, broken spring, and scored seat. A leaky check valve or leaky seals may cause trouble. Excessive heat may cause the valve piston to bind inside the body.

Dirt often causes a hydraulic pressure reducing valve to malfunction. Dirt may clog the small orifices, stopping oil flow through some of the passages.

Maintenance

When servicing a pressure reducing valve, be careful not to lose any of the small parts in disassembling. Clean all of the internal parts carefully, and check the seats, seals, and the fit between the piston and body. If the piston shows any signs of varnish or other foreign materials, clean thoroughly in a good solvent. All parts should be cleaned in a solvent and dried thoroughly. If this valve or any other valve is to be placed in stock after it is reassembled, be sure that the ports are plugged. In reassembling the valve, make certain that the parts are replaced correctly. A manufacturer's parts list for all valves and other components in the system should be available. Replace or repair worn parts. Plating the piston, if worn, is recommended.

Check the valve by using two pressure gauges, place one gauge on the high-pressure side and the other gauge on the low-pressure side. By changing the adjustment needle, determine whether the full range of pressures can be obtained.

SEQUENCE VALVES

Sequence valves are widely used in both hydraulic and pneumatic systems. By using these valves, a second directional control valve can often be eliminated. Hydraulic sequence valves may be direct-acting, direct-operated pilot-type, or the remote-operated pilot-type, the same as for relief valves. Pneumatic sequence valves are generally direct-acting. Fig. 9 shows a pneumatic sequence valve, and Fig. 10 shows a direct-acting hydraulic sequence valve. Fig. 11 shows a direct-operated pilot-type valve that is designed for pipe line mounting. Such valves are also available for subplate mounting. This valve has a built-in check. In Fig. 9, when compressed air enters the port and when the pres-

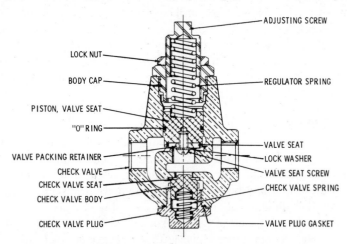

Fig. 9. Pneumatic sequence valve.

sure which acts on the piston overcomes the tension created by the spring, the orifice to the other port is opened; this allows air pressure to flow out the second port. Air pressure can be returned through the valve when the check valve is depressed. The spring tension can be either increased or decreased by movement of the adjusting screw. Pneumatic sequence valves are often used for low-pressure oil service.

The principle of operation of the high-pressure hydraulic sequence valve (see Fig. 10) is similar to the action of the valve shown in Fig. 9. Study the part names of all valves (Figs. 9, 10, and 11).

Pneumatic sequence valves are built for pressures up to 150 psi; hydraulic sequence valves are built for ranges comparable to those of relief valves and reducing valves. Port sizes are also comparable. Pneumatic valves are generally built with threaded

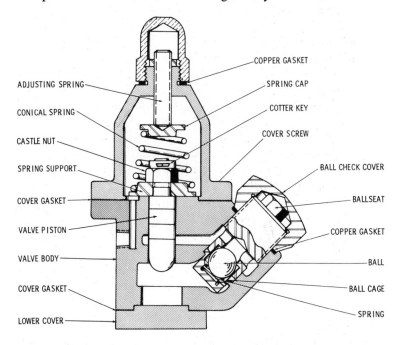

Fig. 10. Direct-acting type of hydraulic sequence valve.

connections, while hydraulic valves may be threaded, flanged, or gasketed.

Hydraulic sequence valves are sometimes used as spring-loaded check valves in complicated systems where several directional control valves are used.

Installation

Both pneumatic and hydraulic sequence valves are usually pipe line mounted, and should be installed so that they can easily be removed from the pipe line. Install sequence valves where

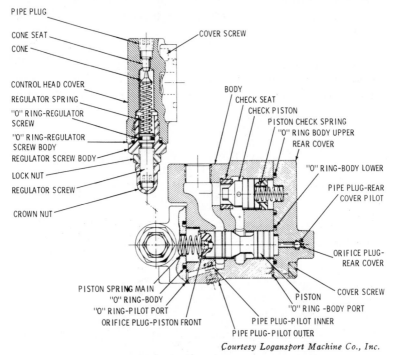

PIPE PLUG
CONE SEAT
CONE
COVER SCREW
CONTROL HEAD COVER
REGULATOR SPRING
"O" RING-REGULATOR SCREW
"O" RING-REGULATOR SCREW BODY
REGULATOR SCREW BODY
LOCK NUT
REGULATOR SCREW
CROWN NUT
BODY
CHECK SEAT
CHECK PISTON
PISTON CHECK SPRING
"O" RING BODY UPPER REAR COVER
"O" RING-BODY LOWER
PIPE PLUG-REAR COVER PILOT
"O" RING-BODY PORT
ORIFICE PLUG-REAR COVER
COVER SCREW
PISTON
PISTON SPRING MAIN
"O" RING-BODY
"O" RING-PILOT PORT
ORIFICE PLUG-PISTON FRONT
"O" RING-BODY PORT
PIPE PLUG-PILOT INNER
PIPE PLUG-PILOT OUTER

Courtesy Logansport Machine Co., Inc.

Fig. 11. Direct-operated pilot-type sequence valve.

they are readily accessible for adjustment and servicing when it is necessary.

Install sequence valves so that they are not subjected to direct blasts of intense heat. Most sequence valves, especially pneumatic valves, contain considerable packing. Fig. 12 and Fig. 13 show where the sequence valves are installed in pneumatic and hydraulic systems.

Causes Of Failure

Most of the same causes of failure in other pressure valves also cause sequence valves to fail—broken springs, dirt, excessive

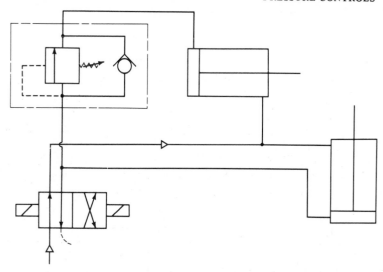

Fig. 12. Use of sequence valve in a pneumatic circuit.

heat, wear, and misapplication. Pneumatic sequence valves also fail from lack of, or improper, lubrication. Improper assembly may cause a valve to fail.

Maintenance

In servicing sequence valves, clean thoroughly, check parts for wear, replace parts where necessary, and assemble properly. If the valves are to be placed in stock after servicing, be sure to plug the ports and to store the valves in a clean, cool, dry place. If valves are to be repainted, be sure to plug the ports and mask the adjusting devices before painting. Keep paint out of all piping threads.

In testing a sequence valve, apply fairly low pressure to the inlet port, and check the outlet port for leakage. Then increase the pressure to the port to determine whether the piston or poppet automatically opens and lets the fluid flow to the outlet port. Check for free flow return by applying low-pressure fluid in the outlet port, to determine whether the check valve opens to permit the fluid to flow from the inlet port. Considerable time and labor may be saved if the sequence valves are tested thoroughly as they are reassembled.

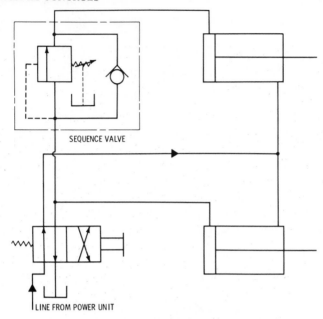

SEQUENCE VALVE

LINE FROM POWER UNIT

Fig. 13. Use of sequence valve in a hydraulic circuit.

UNLOADING VALVES

The unloading valve may be classified as a type of relief valve. It is used to unload pumps, such as in a high-low pumping system, where the large volume low-pressure pump is unloaded automatically when a certain pressure is reached. It also can be used for other unloading applications where it is necessary to dump the oil pressure during stand-by periods when the pump is not operating.

Unloading valves are remotely operated. Fig. 14 shows the pilot connection to this valve. Fig. 15 shows how the unloading valve is used in a system where a high-low pumping device is employed.

The principle of operation is that the pilot pressure causes the piston to move against the spring tension; thus the inlet port becomes connected to the exhaust port, which directs the fluid back to the sump or reservoir. Unloading valves are built in sizes and pressure ranges that are similar to relief valves. Installation,

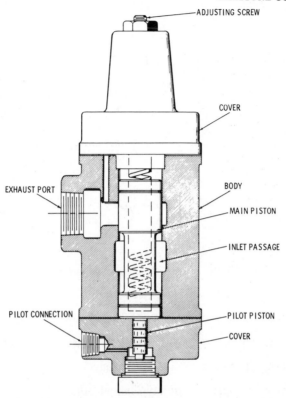

ADJUSTING SCREW

COVER

EXHAUST PORT

BODY

MAIN PISTON

INLET PASSAGE

PILOT CONNECTION

PILOT PISTON

COVER

Fig. 14. Showing pilot connection in an unloading type of valve.

causes of failure, and maintenance procedures are similar to those for the relief valve.

OTHER TYPES OF PRESSURE CONTROLS

Other types of pressure controls, some of which are modifications of the ones discussed previously, are used in fluid power systems. These are: pressure switches, both pneumatic and hydraulic; counterbalance valves; pressure relay valves; overload relief valves; and pressure-regulating unloading valves. Pressure switches are electric switches that are controlled by fluid pressure.

287

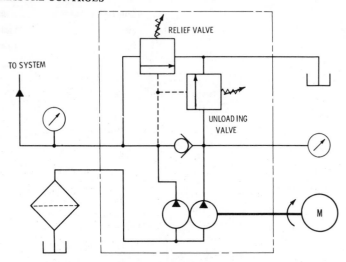

Fig. 15. Use of an unloading valve in a circuit.

Review Questions

1. Explain the operation of a typical relief valve.

2. Explain the operation of a typical pressure reducing valve.

3. Explain the operation of a typical sequence valve.

4. Explain the operation of a typical unloading valve.

5. Why are two relief valves sometimes used in a circuit?

6. List the factors that should be considered when installing a relief valve.

7. What conditions may cause a relief valve to fail in operation?

8. In a pressure reducing valve, what is the chief difference between the direct-acting type and the direct-operated pilot-type valve?

9. What conditions may cause a pressure reducing valve to fail in operation?

10. What conditions may cause an unloading valve to fail in operation?

CHAPTER 14

Rotating and Nonrotating Cylinders

The function, installation, and maintenance of rotating and nonrotating hydraulic and pneumatic cylinders are discussed in this chapter. The cylinder is the component of the hydraulic or pneumatic system that receives fluid, under pressure, from a supply line. The fluid in the cylinder acts on a piston to do work in a linear direction. The work that is performed is the product of the fluid pressure and the area of the cylinder bore, as illustrated in Fig. 1. The speed or rate of doing work depends on the quantity of fluid delivered into the cylinder.

Applications of the nonrotating type of cylinder are shown in

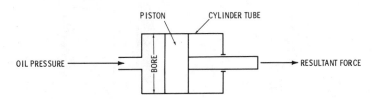

Fig. 1. Schematic diagram of a power cylinder.

Fig. 2. A typical application for the rotating cylinder, as used on an automatic lathe, is shown in Fig. 3.

TYPES OF CYLINDERS

The following terms should become familiar: (1) In the double-acting nonrotating and rotating cylinders, the fluid supply pres-

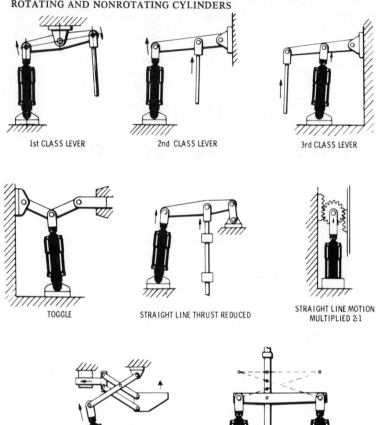

1st CLASS LEVER

2nd CLASS LEVER

3rd CLASS LEVER

TOGGLE

STRAIGHT LINE THRUST REDUCED

STRAIGHT LINE MOTION
MULTIPLIED 2:1

STRAIGHT LINE MOTION IN TWO DIRECTIONS

4 POSITIVE POSITIONS WITH TWO CYLINDERS

Fig. 2. Applications for the

sure can be applied to either side of the piston, so that work can be done in either direction; and (2) In the single-acting nonrotating and rotating cylinders, the fluid supply pressure is applied only to one side of the piston. In the spring-return type of cylinder, the piston is returned to its starting position by spring action after the fluid pressure has been released from the piston. The direction of work depends on the arrangement of the piston seal.

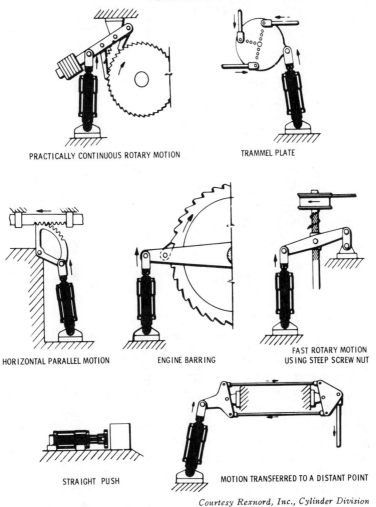

PRACTICALLY CONTINUOUS ROTARY MOTION

TRAMMEL PLATE

HORIZONTAL PARALLEL MOTION

ENGINE BARRING

FAST ROTARY MOTION
USING STEEP SCREW NUT

STRAIGHT PUSH

MOTION TRANSFERRED TO A DISTANT POINT

Courtesy Rexnord, Inc., Cylinder Division

nonrotating type of cylinder.

The plunger or ram-type cylinder, nonrotating type only, is another type of single-acting cylinder, but it contains no piston. The ram-type cylinder has only one diameter, whereas the piston-type cylinder has more than one diameter. In a single-acting cylinder, either the piston or the ram may be returned by gravity or some mechanical means, if a spring is not used.

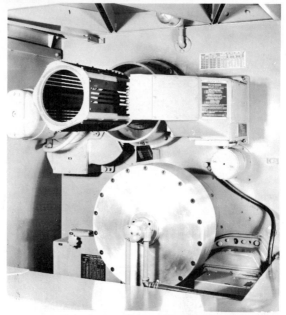

Fig. 3. Rotating air cylinder used on an automatic lathe.

These types of nonrotating cylinders are illustrated in Fig. 4. A single-acting spring-return type of cylinder is shown in Fig. 5.

In the "double-end" cylinder, the piston rod extends through both ends of the cylinder, and pressure may be applied to either side of the piston. In the hollow-center cylinder (rotating type only), the center portions of the piston and distributor are hollow, which allows bar stock or coolant to pass through the center.

In a "tandem" type of cylinder, two or more pistons are mounted on a common piston rod. This allows more force to be exerted by the end of the piston rod for a given cylinder diameter. This type of cylinder is often used in places where the diameter of the space in which the cylinder can be located is at a premium. Applications for this type of cylinder are: cushioning on dies, workholder devices, and lathe equipment.

A "duplex" type of cylinder contains two pistons and two piston rods; one of the piston rods works inside the other piston rod. These cylinders are generally used where a sequence opera-

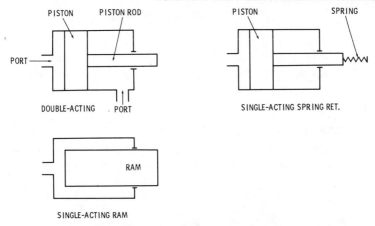

Fig. 4. Schematic diagrams of nonrotating cylinders: (A) double-acting; (B) single-acting, spring-return; and (C) single-acting ram-type.

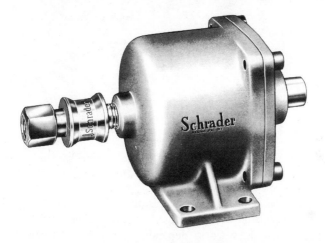

Fig. 5. Single-acting spring-return type of cylinder.

tion is required; one piston rod clamps the workpiece, and the other piston rod performs another operation. Duplex cylinders are more expensive, and they are more difficult to service because of their complicated design. The double-end, tandem, and duplex types of cylinders are usually called "specials," and they are not commonly manufactured as catalogued items.

NONROTATING CYLINDERS

A cutaway view of a double-acting cylinder with a cushioning device is shown in Fig. 6. Another view of a double-acting cylinder is shown in Fig. 7. The cylinder in Fig. 6 is used for air service. The cylinder in Fig. 7 can be used for either air service or low-pressure hydraulic service.

Part Names

The names of the parts of the cylinder and their functions should be studied to understand how the cylinder operates. A

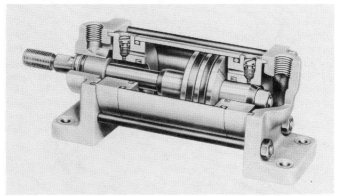

Fig. 6. Double-acting cylinder with cushioning device.

high-pressure hydraulic cylinder with the names of the parts is illustrated in Fig. 8. The important parts are as follows:

1. *Piston rod.* One end of the piston rod is connected to the piston (see Fig. 8), and the other end is connected to a device which does work, depending on the requirements of the job. Two different types of rods are shown. One type is shown as the standard-size rod; the other type is a two-to-one rod, in which the cross-sectional area of the rod is approximately one-half the area of the cylinder bore. The latter type of rod is seldom used in air cylinders. Piston rods must be extremely durable, since they are often subjected to grinding compounds and other adverse conditions. Piston rods are made of a good grade of steel that is ground

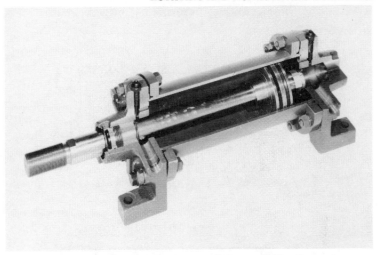

Fig. 7. Double-acting cylinder for pneumatic or hydraulic service.

and polished to an extremely smooth finish, and they may be hardened and chromeplated to resist wear. Stainless steel is often used to resist corrosion.

2. *Rod wiper.* A rod wiper (made of durable synthetic material) is used to clean the piston rod as it is retracted into the cylinder. All foreign matter on the rod must be removed before the rod is drawn back into the packing. A metallic scraper (Fig. 9) is sometimes needed to remove severe residues.

3. *Cylinder covers.* Each cylinder has two covers—the front or rod-end cover and the blank or blind-end cover. Sometimes the blind-end cover is part of the cylinder body or tube. The functions of cylinder covers are: to seal the ends of the cylinder tubes; to provide a method of mounting; to provide a housing for seals, rod bearing, and rod packing (front cover); to provide for ports of entry for the fluid; to absorb the impact of the piston; and to provide room for the cushioning arrangement.

   Cylinder covers may be made of iron, steel, bronze, or aluminum. Some of the newer designs are made of steel bar stock. The dimensions of covers are well standardized.

4. *Cylinder tube.* The cylinder tube may be made of cold-

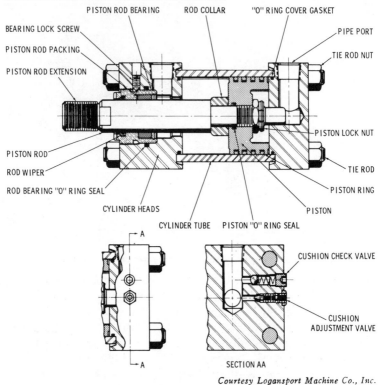

Courtesy *Logansport Machine Co., Inc.*

Fig. 8. Heavy-duty hydraulic cylinder.

drawn seamless steel, brass, or aluminum tubing, held to close tolerance and honed to an extremely smooth finish. The sealing action of the piston depends largely on the finish obtained in the cylinder tube. Cylinder tubes may also be made of centrifugal cast iron, cast steel, or centrifugal cast bronze, the latter being used mostly for water service. The wall thickness of centrifugal cast materials is usually much heavier than seamless tubes. The covers are often fastened to the cylinder tube with cover screws, as shown in Fig. 10.

5. *Piston assembly*. This assembly functions in a manner similar to the pistons in an automobile. The piston must fit closely to the cylinder wall to provide a suitable bearing

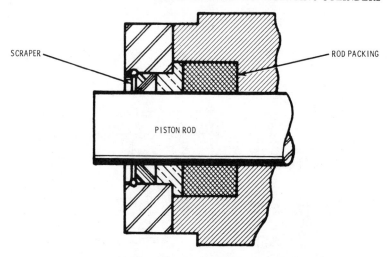

Fig. 9. Illustrating a metallic scraper, used to remove residues.

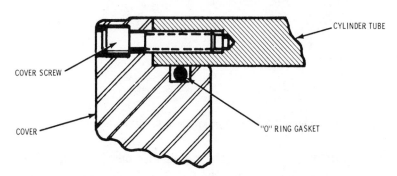

Fig. 10. Cover screws hold the cover to the cylinder tube.

and to eliminate any possibility of extrusion of synthetic seals. Since the function of the piston is to act as a bearing, it must be made of a material that will not score the cylinder tube. A high tensile strength cast iron with a high surface finish has been found to perform exceedingly well. Although Figs. 6 and 7 show a piston with synthetic packing, one of the most common types of pistons for hydraulic service is shown in Fig. 8 and in Fig. 11. This piston uses automotive-type rings, which may be made of

297

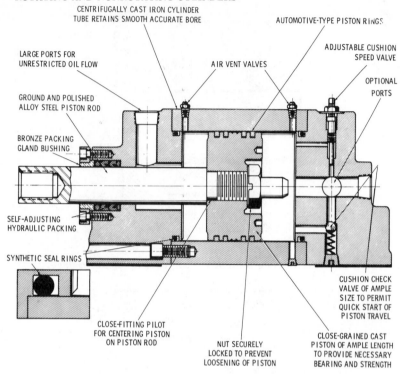

Fig. 11. Automotive-type rings are used in this hydraulic cylinder.

cast iron or bronze. The piston with automotive rings has a relief at each end, so that any small particles of dirt which may get into the system will not spring the end land and cause the piston rings to be crushed or frozen. This is a design feature that was perfected after many years of experimentation. Other types of sealing means used in pistons are chevron packings, cup packings, and "O" rings.

6. *Piston lock nut.* A locking-type nut prevents the piston from coming loose on the piston rod; it is compact and secure. A lock nut eliminates the need for setscrews, pins, and other locking means, which often work loose and drop into the cylinder to cause considerable damage.

7. *Tie rods.* Tie rods are used to hold the cylinder together. They must be strong enough to absorb the shock loads as the piston contacts the cylinder cover. Tie rods are made of alloy steel, and they must have an ample safety factor. Although the design shown in Fig. 8 uses tie rods, other designs use such means as through-bolts (Fig. 7), threaded tubes, snap rings, welded construction, cover screws tapped into cylinder flanges, or tubes to hold the cylinder together. An advantage of the tie-rod construction is that the tie rods have a certain amount of elasticity, which relieves shock.

8. *Cushion collar and nose.* The function of the cushion collar and nose is to alleviate shock as the piston approaches the cylinder covers. It not only alleviates the shock to the piston and the covers but also eliminates the sudden stopping of an object that is connected to the end of the piston rod. For example, if a basket containing dishes is pushed and then stopped suddenly, the dishes will spill and break. The cushioning device eliminates this. The

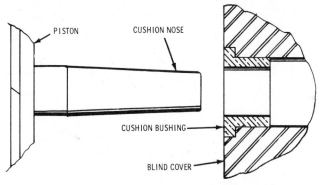

Fig. 12. Illustrating use of a tapered cushion nose.

cushion device on the rod-end side of the piston of the cylinder is known as the cushion collar, and the device on the blind side is called the cushion nose. On many applications, standard-length cushions (which have an effective length of about one inch) are not always long enough. In these instances, extra-long cushions should be installed. Adding length to the cushion on the blind end of

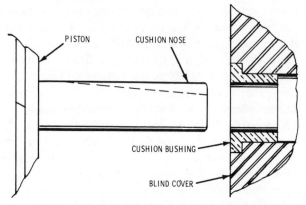

Fig. 13. Use of a vee-groove in a cushion nose.

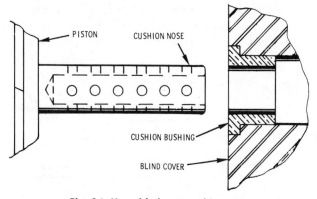

Fig. 14. Use of holes in cushion nose.

the cylinder usually does not present any problem, since the blind-end cover of the cylinder can usually be drilled through and tapped. Then a pipe nipple with a cap can be screwed into the cover to receive the longer cushion nose. The rod end presents a more difficult problem, since some type of spacer must be used inside the cylinder tube to take care of the added cushion length. This shortens the stroke of the cylinder, or if the same stroke is required, the overall length of the cylinder becomes greater.

Longer cushion lengths are often required on cylinders moving a large mass on wheels, bearings, or other free-moving means at high speed. Some of the cushioning designs that have proved very useful are shown in Figs. 12, 13, and 14. Cushions six to eight inches long are often necessary. A gradual taper on the cushion nose is shown in Fig. 12 which, as it enters the cushion bushing, gradually closes the recess and traps the fluid. This fluid must then flow through a preset orifice. In Fig. 13, the cushion nose, which is only a few ten-thousandths of an inch smaller in diameter than the cushion bushing, enters the bushing, but the vee-groove in the nose allows the fluid to escape, until the groove is gradually closed off. Then the fluid must escape through the preset orifice. In Fig. 14, the holes are closed off as the cushion nose penetrates farther into the bushing until all the holes are closed off; then the fluid must escape through the preset orifice. The advantage of these types of cushions is that the speed is gradually reduced, instead of being reduced abruptly.

9. *Cushion adjustment valve.* This valve works with the cushion nose and collar and is a part of the entire cushioning arrangement. When the trapped fluid (this takes place after the cushion closes the bushing) is metered out, it must pass by the needle of the cushion adjustment valve which forms the size of the exhausting orifice. The trapped fluid cannot flow past the ball check, as it is blocked in that direction. Although designs are also made with a fixed exhaust orifice, the chief advantage of the adjustable orifice is that it can be changed for various loadings and operating pressures.

In order to allow the fluid to start the piston quickly in motion at the beginning of the stroke, the ball check is opened. The fluid flows freely through the check into the cylinder, until the cushion nose or collar clears the cushion bushing. If it were not for the check valve, much time would be consumed in waiting for the fluid to flow back into the cylinder through the fixed orifice of the cushion adjustment valve.

Some recent developments have eliminated the ball check, and have made use of a synthetic check (see Fig. 6) located in the cylinder cover so that, as the cushion collar or nose enters the cover, a positive seal is effected. When fluid pressure is applied in the opposite direction, the seal collapses and allows the fluid to flow freely to the piston.

10. *Piston rod bearing.* The piston rod bearing not only houses the rod packing but also acts as a bearing and as a guide for the piston rod. Rod bearings are made of a good-quality bearing bronze or cast iron.

11. *Rod packing.* The rod packing seals the piston rod, so that the fluid cannot escape around the rod. Piston rod packings are of various designs, such as chevron, blocked vees, "O" rings, quad-ring, and *Sea* ring. Different types of materials, such as synthetic rubber, leather, *Teflon*, and nylon, are used in packings, depending on the application. Impregnated materials are also used.

12. *Cover gaskets.* Cover gaskets act as a seal between the cylinder cover and the cylinder tube. With the "O" ring placed as shown in Fig. 8, a perfect seal is formed, which is practically indestructible. The "O"-ring seal becomes tighter as the pressure is increased. Note that "O" rings are used for seals in many places in this cylinder.

The major parts of a nonrotating cylinder have been discussed. Parts vary with the different manufacturers, but, in general, nearly all cylinders have the aforementioned parts.

Forces in a Nonrotating Cylinder

Before discussing actual installation of nonrotating cylinders, the great amount of force that they can develop and their requirements when mounted in a system should be realized (Table 1).

The table is based on the simple formula:

$$F = PA; \text{ and } F = P(A - A_1)$$

in which;

F is force developed (theoretical, not including friction).

P is supply pressure (psi).

A is area of cylinder bore (sq. in.).

A_1 is area of cross section of piston rod (sq. in.).

Table 1. Force Developed By Nonrotating Cylinders

Cylin-der Bore	Area of Cyl. Bore Rod	Rod Dia.	At 80 psi		At 100 psi		At 500 psi		At 1000 psi		At 1500 psi	
			Push#	Pull#	Push#	Pull#	Push#	Pull#	Push#	Pull#	Push#	Pull#
2	3.142	1	251	188	314	235	1571	1178	3142	2357	4713	3535
2½	4.909	1⅜	392	273	490	342	2454	1712	4909	3424	7363	5136
3	7.069	1⅜	565	443	706	558	3534	2792	7069	5584	10603	8376
3½	9.621	1⅜	769	650	962	813	4810	4068	9621	8136	14431	12204
4	12.566	1¾	1005	812	1256	1016	6283	5080	12566	10164	18849	15241
5	19.635	2	1570	1319	1963	1649	9817	8246	19635	16493	29457	24739
6	28.274	2	2261	2010	2827	2513	4137	12566	28274	25132	42411	37698
7	38.485	2½	3078	2686	3848	3357	19242	16788	38485	33576	57727	50364
8	50.265	2½	4021	3628	5026	4535	25132	22678	50265	45356	75397	68034

The formula, $F = PA,$ is used to find the force developed when fluid pressure is applied at the blind end of the cylinder (the end opposite the piston rod), and the formula, $F = P(A - A_1)$, is used to find the force created when fluid pressure is applied to the rod end of the cylinder. Note that when the same operating pressure is applied either to the fluid end or to the rod

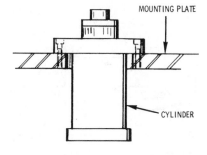

Fig. 15. Reverse-flange mounting for press applications.

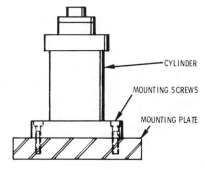

Fig. 16. Cylinder adaptable for blind-flange mounting in the base of a machine.

303

end of a cylinder, the higher force is always exerted by the blind end, because it has a greater area.

Nonrotating cylinders are designed for the various pressure ranges, and should be used for applications within those ranges. For standard cylinders, these ranges are designated in pounds per square inch of operating pressure: (1) for air, the ranges are 0-150 and 0-250; and (2) for oil, the ranges are 0-150, 0-750, 0-1500, 0-2000, and 0-3000. Special cylinders are designed to operate as high as 10,000 psi, but these applications are rare.

Installation

The different cylinder mounting styles should be studied to determine how they can be installed for best results. It should be remembered that installation plays an important part in the performance of a cylinder. All mountings must be fastened securely.

Flange-Mounted—The front-flange mounted cylinder is well adapted for "pulling" applications. In this type of mounting, the front cover bears against the mounting plate, thus relieving the

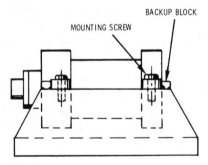

Fig. 17. Cylinder with mounting lugs at the center line.

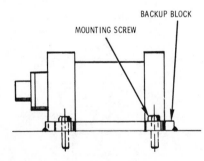

Fig. 18. Illustrating a commonly used foot mounting style.

pressure on the mounting screws. On press applications, reverse-flange mounting (Fig. 15) is preferable, because the strain on the mounting screws is relieved. The blind-end or flange-mounting type (Fig. 16) is especially desirable where it is necessary to mount the cylinder in the base of a machine. Some machines can use only this type of cylinder.

Center-Line Mounted—An advantage of the center-line mounted cylinder (Fig. 17) is that the mounting feet or lugs are in a direct line with the center line of the thrust. The mounting feet should be keyed, in order to eliminate the full thrust on the mounting bolts.

Foot-Mounted—The foot-mounted style (Fig. 18) is one of the most commonly used methods of mounting nonrotating cylinders. Since the mounting lugs are located in a different plane from that of the center line of the thrust, some torque is created. Here again, the mounting lugs should be keyed to reduce the thrust on the mounting bolts. In mounting this type of cylinder, make sure that the piston rod has ample support, especially on cylinders that have long strokes. The same type of guide should be placed on the end of the rod to prevent sagging, which causes wear on the rod bearing or damage to the piston or cylinder walls. Extra-long cylinders should be supported in the center of the cylinder tube (Fig. 19) to reduce sagging of the tube. This support should be in perfect alignment with the mountings at the front and rear of the machine.

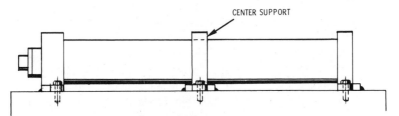

CENTER SUPPORT

Fig. 19. Long-stroke foot-mounted cylinders should be supported at the center of the cylinder.

Loading is always a problem with nonrotating cylinders. These cylinders should be loaded concentrically wherever possible. If this is impossible, provisions should be made for eccentric loading. One method of compensating for eccentric loading is to use heavy guide rods and bearings, as shown in Fig. 20.

305

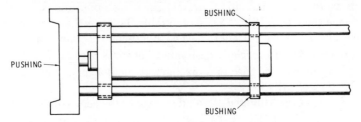

Fig. 20. Method of providing for eccentric loads.

Causes of Failure

The major causes of nonrotating cylinder failures are:

1. *Dirt.* The one problem that causes more failures than any
   thing else is—dirt. Dirt may lodge between the piston and
   the cylinder tube, and score the cylinder tube; the result
   is that the piston seal becomes defective, and excessive
   leakage past the piston occurs. If the scoring is deep enough.
   it may cause the piston to freeze to the tube. Dirt settling
   on the piston rod may score the rod as the rod is drawn
   back into the cover; here again, leaks develop.
2. *Heat.* This causes deterioration of the packings, causing
   packing and gasket leaks. Temperature at the cylinder should
   not exceed 140°F. If the cylinders are subject to excessive
   external heat, some type of heat-resistant shield should be
   provided. Also, it is wise to install heat-resistant packings.
   These packings are now available for temperatures up to
   500°F.
3. *Misapplication.* A large percentage of cylinder failure can
   be attributed to misapplication, rather than to faulty design of a cylinder. For example, a cylinder with cast iron
   covers should not be used in applications involving high
   shock impact and eccentric loads. In some heavy-duty applications, a cylinder may fail in a few days, whereas the
   same cylinder used on a medium-duty application may give
   satisfactory service for many years.
4. Misalignment, side thrust, or eccentric loading that is not
   supported properly may cause cylinder failure. These factors may be noticed first by excessive wear on one side of

the piston rod, by rod packing leaks, or by wear on one side of the rod bearing. More serious damage may result from such conditions as a bent piston rod, a broken packing gland, a scored cylinder tube, broken cylinder covers, or a broken piston.

5. Faulty mountings are a cause of cylinder failure. If the cylinder is not mounted securely or if the mounting is not of sufficient strength to withstand the load produced by the cylinder, the cylinder may break loose and damage not only the mounting but also itself and the application with which it is connected.

6. In a pneumatic cylinder, failure is often caused by lack of lubrication. Make certain that proper lubricant in sufficient quantities is fed to the pneumatic cylinders.

Repair And Maintenance

Some suggestions for dismantling, repairing, and assembling a nonrotating cylinder are:

1. Dismantle the cylinder in a clean location. If it is a hydraulic cylinder, drain all the oil from the cylinder. Do not try to dismantle a cylinder when there is pressure to the cylinder.

2. Clean each part. If the cylinder is to be dismantled for any length of time, coat the metal parts that are to be reused with a good preservative and store in protected storage.

3. Check the piston rod for straightness. If the rod is bent, place it on vee blocks in a press, and carefully straighten it.

4. Examine the piston rod for scratches, scores, indentations, and other blemishes. If these blemishes are not too deep, remove them with a fine emery cloth. If it is necessary to grind the rod, it is suggested that it be chrome plated afterwards to restore the diameter to its original size.

5. Examine cover bushings and cushion bushings for wear and finish. If they are not in first-class condition, they should be replaced.

6. If the cylinder tube is damaged, either repair or replace it. Deep scores are very difficult to repair. It may be necessary to chrome plate the tube.

7. In reassembling a cylinder, it is suggested that all the seals

and gaskets be replaced. When metal piston rings are used, check the cylinder manufacturer's specifications for gap clearance. If synthetic or leather seals are used on the piston, be extremely careful in placing the piston in the tube, so that sealing surfaces are not damaged. Use a light grease on packings when installing them; this makes them much easier to assemble.

8. If, for any reason, a metal piston with rings is to be replaced, be sure to grind the piston concentrically with the piston rod after it has been assembled to the rod. Grind the piston to fit the tube closely.

9. Cylinders that have foot-mounted covers should be assembled on a surface plate, and the mounting pads of both covers must make full contact with the surface plate; otherwise, a binding action may occur when it is mounted, or a mounting foot may be broken.

10. Always tighten the cover bolts evenly. If "O"-ring or quad-ring gaskets are used to seal the tube and cover, tension on the cover screws can be reduced to a minimum.

11. After a cylinder has been completely assembled, test it at low operating pressure to make certain that the piston and rod are moving freely and that they are not being scored or bound. Then increase the pressure to its full operating range, and check for both internal and external leakage. To check internal leakage, place fluid pressure in the blind-

Courtesy Rexnord, Inc., Cylinder Division

Fig. 21. Pneumatic cylinder with protective boot on piston rod.

308

end cylinder port, and force the piston to the rod end; then check the amount of fluid that comes out of the rod-end cover port. Then place fluid pressure in the rod-end cylinder port, move the piston to the blind-end cover, and check leakage at this position. If leakage is to be checked at other positions of the cylinder, block the ports at these positions by external means and check as above. If excessive leakage occurs, it is necessary to disassemble and make corrections. If the piston is sealed with synthetic or leather seals, the leakage should be practically nil. If metal piston rings are used in a hydraulic cylinder, the leakage will vary with the operating pressure, the cylinder bore, and the oil temperature and viscosity.

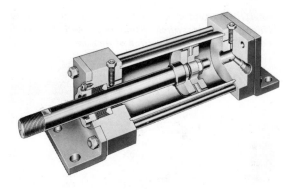

Courtesy Logansport Machine Co., Inc.

Fig. 22. Cutaway view of pneumatic cylinder with vee-type rod seal and cup-type piston seals.

12. When returning a cylinder to a machine or fixture, make certain that it is securely mounted. Remember that these cylinders are capable of delivering a great amount of force.

Examples of nonrotating cylinders are shown in Figs. 21, 22, and 23. Cylinder bores range from less than 1 inch in diameter to more than 30 inches in diameter. Strokes are available from less than 1 inch to more than 30 feet.

A   B

Courtesy The Sheffer Corporation

Fig. 23. Limit switch-actuated hydraulic cylinder. Air limit switches are also available. (A) Limit switch eliminates need for expensive trip mechanisms. (B) Cross section of limit switch actuator.

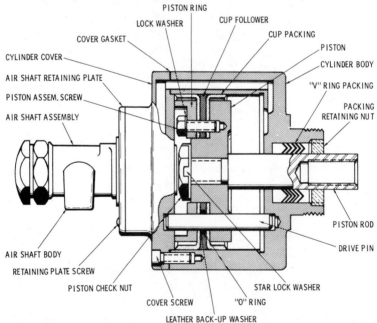

Fig. 24. Rotating pneumatic cylinder.

ROTATING CYLINDERS

The names of the parts and their functions should be studied to understand how the rotating cylinder operates. A rotating pneumatic cylinder and a rotating hydraulic cylinder are diagrammed in Fig. 24 and Fig. 25, respectively.

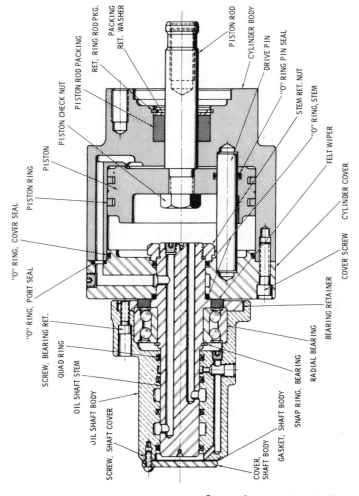

Courtesy Logansport Machine Co., Inc.

Fig. 25. Rotating hydraulic cylinder.

311

Part Names

The important parts of rotating cylinders are:

1. *Cylinder body.* The cylinder body is usually made of either cast iron or cast aluminum, and it must be pressure tight. This body is usually machined on both the inside and outside surfaces—on the I.D. (inside diameter) to provide a smooth surface for the packing and on the outside for the sake of appearance. The body contains air passages, so that the fluid can be directed to the front side of the piston. The body also contains the housing for the rod packing, and provides a bearing for the piston rod. The open end of the body has a number of tapped holes to receive the cover bolts. The closed end of the body is designed with an adaptation which may follow either American Standard or a manufacturer's specifications. The cylinder body may also anchor one end of the drive pin.

   Where the cylinders are designed with longer than standard stroke, the body is often made in two sections—the cylinder tube and the rod-end cover. This helps reduce porosity, which is often a problem on long-stroke rotating cylinders when the cylinder body is constructed in a single piece.

2. *Cylinder cover.* The cylinder cover encloses the cylinder, and is fastened to the cylinder body by cover screws. The cylinder cover is usually made of the same material as the cylinder body—cast iron or aluminum. The cylinder cover carries the air shaft or oil shaft assembly. The cover may also anchor one end of the drive pin. The cylinder cover carries a fluid passage which connects to the fluid passage in the cylinder body.

3. *Piston rod.* The piston rod is the connector between the piston and the driven means. It is made of a ground and polished alloy steel. The finish must be such that the packing will last. The end of the piston rod is usually tapped with female threads; however, some applications may require male threads.

4. *Piston assembly.* The piston asssembly of the air cylinder shown in Fig. 24 uses cup packings for seals; however, the piston assembly of the hydraulic cylinder in Fig. 25

uses an automotive-type piston ring. Other piston designs make use of various types of seals, such as "V" packing and blocked vees. The piston may be made of either cast iron or aluminum. The ring which backs up the packing in this particular cylinder is made of a synthetic material which acts as a bearing as well as a support for the packing. Although this piston assembly is composed of parts known as the piston, piston ring, piston follower, piston assembly, screws, and piston packings or seals, it is conceivable that a piston assembly could be designed with only two parts—a piston and a piston seal (Fig. 26). The piston has three main functions. One function is to provide

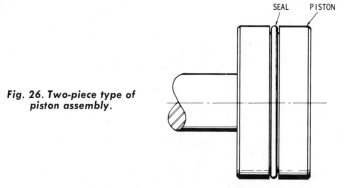

SEAL   PISTON

Fig. 26. Two-piece type of piston assembly.

an area over which the fluid pressure acts; thus it develops a force to act on a load. The second function is to provide a seal which prevents the fluid from leaking or escaping to the exhaust side. The third function is to act as a guide or bearing.

5. *Drive pins.* The function of the drive pins is to keep the piston from rotating in relation to the cylinder body. Since brakes are provided on many machines with rotating spindles, there is a need for drive pins. Without them, a sudden stop of the spindle may cause the piston to rotate within the body, causing the piston rod to become loosened or disconnected from its connecting part. This often causes considerable trouble and consumes considerable time to correct.

6. *Rod packing.* The rod packing on a rotating cylinder is extremely important, as most of the work is usually done on the instroke, which throws fluid pressure on the rod packings most of the time. Any leakage past the packings greatly reduces the efficiency on the instroke. Packings on rotating cylinders must be very durable as it is quite a job to make

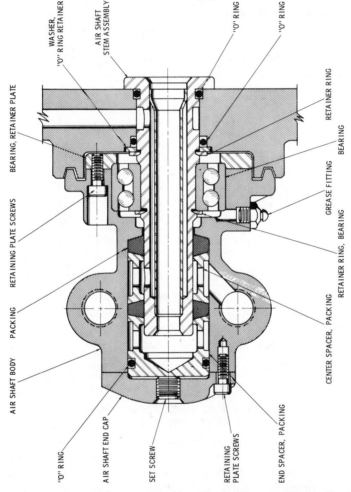

Courtesy Logansport Machine Co., Inc.

Fig. 27. Air shaft assembly with hollow tube in air shaft.

a packing change, because these cylinders are usually mounted on a machine spindle.

Packings may be of several designs, such as chevron, *Sea* rings, and hat. Materials used are synthetic rubber, impregnated leather, *Teflon*, and others.

7. *Cover gasket.* The cover gasket seals the cylinder body and the cylinder cover. It contains holes for the cover screws and for the fluid passages between the body and the cover. The cover gasket may be made of a thin sheet of gasket material, but it must be strong enough to keep the fluid from escaping.

8. *The air shaft or oil shaft stem.* The air shaft or oil shaft stem plays an important part in the action of the rotating cylinder. It must be sturdy enough to provide ample support to the remainder of the air shaft and oil shaft assemblies and to withstand external forces that are caused by the weight of flexible hoses and pipe connectors. The air shaft or oil shaft stem is made of a hardened alloy steel ground to a high finish, so that the packing will have a smooth bearing surface.

Although the air shaft shown in Fig. 27 contains a hollow tube inside it for providing air passages without extensive drilling, other designs make use of small drilled holes for carrying the air (Fig. 28). The air shaft assembly is

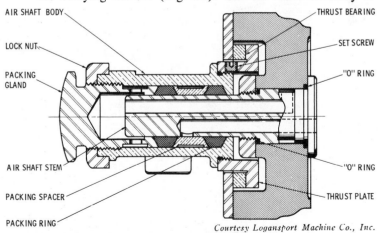

Courtesy Logansport Machine Co., Inc.

Fig. 28. Air shaft with drilled holes.

often referred to as an air distributor, and an oil shaft assembly is referred to as an oil distributor.

9. *Distributor body.* The functions of the distributor body are to act as a housing for the shaft packing, to house the means of lubrication, to provide ports of entry, and to help retain the bearing. The distributor body may be made of such materials as cast iron, cast aluminum, or cast bronze.

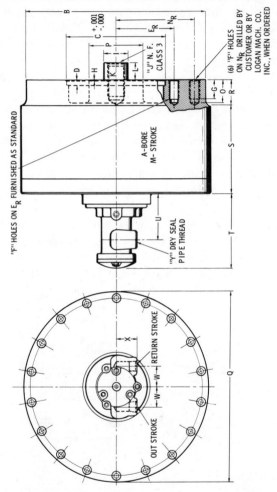

Fig. 29. Mountings for pneumatic rotating cylinders.

It should be so arranged that it will dissipate as much heat as possible from the packings and shaft. The distributor body must be able to withstand the strain of pipe lines and connectors.

Where hollow air shafts or oil shafts are used, it is often necessary to use water-cooled distributor bodies to cool the shaft and packing. These bodies often contain large-cored passages to effect a better cooling action.

10. *Shaft packing.* The shaft packing must create as little friction as possible, yet provide an effective seal. This is a big order. Shafts may revolve at up to 5000 rpm, and higher, and the seal must be able to function under these conditions. Such materials as shredded lead, lead and graphite, asbestos, and other materials are used to combat the heat condition. Various shapes of packings are used to perform the job, such as a formed-wedge, vee, and hat types.

11. *Bearing.* Some type of bearing is required to provide support between the shaft body and the shaft stem. Various types of bearings are used, depending on the shaft design, such as ball bearings, sleeve bearings, and thrust bearings. These bearings must be able to withstand considerable heat. Lubrication to the bearings is very important.

A number of minor parts are used in a rotating cylinder. However, they are important in the functioning of the cylinder. Some of these parts are: retainer rings, spacers, "O"-ring gaskets, and screws. Incorrect reassembly of these minor parts of a rotating cylinder or failure to replace worn or damaged seals, gaskets, etc. may cause the cylinder to malfunction or fail to operate properly, resulting in increased maintenance and repair costs.

Installation

Before actual installation of a rotating cylinder, the pressure chart (see Table 1) should be studied, in order to visualize the amount of force that these cylinders can develop. It is seldom that rotating air cylinders are operated at more than 90 pounds of air pressure. Hydraulic rotating cylinders may be operated as high as 1000 psi.

The installation of a rotating cylinder is very important, because the cylinder may be subjected to high revolving speeds. Rotating cylinders, as stated previously, are constructed with the mounting as a part of the cylinder body. On the small-diameter cylinders, the mounting is usually threaded; on the larger sizes, there are a number of tapped mounting holes (Fig. 29).

Rotating cylinders do not mount directly onto a machine spindle; they are mounted on an adapter, which has been mounted on a machine spindle. This is because each manufacturer of lathes and other machines with rotating spindles uses different end designs on the cylinder end of the spindle. A manufacturer may produce a dozen or more different sizes of spindles, depending on the range of sizes of his machine, each spindle end having different dimensions. The cylinder adapter is usually made of the same kind of material as the rotating cylinder body. If an aluminum cylinder is used, an aluminum adapter is also used.

It is important that the adapter should be properly mounted before the cylinder is installed. The adapter should bottom against the end of the spindle. Then it should be locked securely. Depending on the adapter design, there are several ways of accomplishing the locking means. On a split-type adapter, the locking action is accomplished by tightening the locking screws. On a threaded-type adapter, bronze plugs are forced inward against the threads on the spindle, and locked in place with setscrews. Two plugs are generally used, spaced at 90° apart (Fig. 30). After the adapter is securely locked in place, the pilot on the end that adapts to the cylinder should be checked with an indicator for runout. If the runout is more than 0.002 inch, the adapter should be corrected. Since nearly all rotating cylinders need a draw bar to make a connection to the mechanism which it operates (Fig. 31), the draw bar is screwed into the end of the piston rod of the cylinder. The cylinder is then mounted onto the cylinder adapter with mounting screws. Make certain that the adapting surfaces on the cylinder and adapter are clean and free of any burrs, as these may cause trouble; also, select mounting screws that are short enough that they will not bottom in the mounting holes in the cylinder. Otherwise, the cylinder will not be tight enough on the adapter. After the cylinder has been mounted, check O.D. (outside diameter) of the cylinder for runout with the indicator. If the

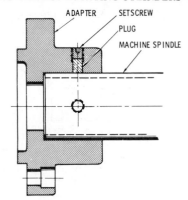

Fig. 30. Adapter for a rotating type of cylinder.

runout is more than 0.003 inch, the cylinder mounting should be corrected.

After the cylinder is mounted, the piping should be connected to the distributor body. As shown in Fig. 31, short lengths of flexible hose should be run from the air body to the rigid piping. This reduces the thrust on the air shaft body.

Causes of Failure

The major causes of rotating cylinder failure are:

1. *Lack of lubrication.* A large number of rotating air cylinder failures can be attributed to a lack of lubrication. In the cylinder itself, lack of lubrication causes both piston packing and rod packing failure due to increase in friction. Lack of lubrication in the air shaft assembly causes rapid packing wear, heat, and bearing failure. Lack of lubrication to the bearings causes a hydraulic rotating cylinder to fail.
2. *Dirt.* Dirt is very detrimental to the interior of a rotating cylinder. It causes packing and bearing failure, scoring of the piston rod, shaft stem, and cylinder body, and can also foul the passages within the cylinder.
3. *Misapplication.* Rotating cylinders are built for various speeds. If the cylinders are used at speeds far in excess of that for which they are designed, failure can be expected. This usually results in packing and bearing failure.
4. *Poor installation.* If the cylinder is not solidly mounted

319

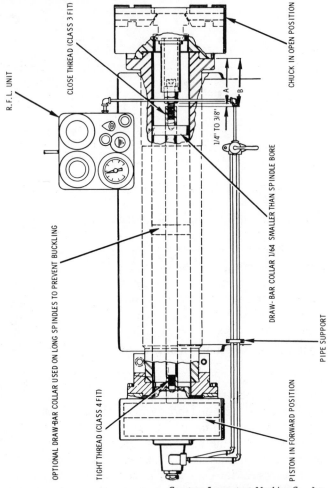

Courtesy Logansport Machine Co., Inc.

Fig. 31. Installation of power-operated chucking equipment on a machine.

or if it is not mounted concentric with the spindle, considerable trouble can result. This causes a whipping action on the back end of the cylinder, which soon causes leakage of the shaft packing and breakdown of the bearings. If the piping to the distributor body causes any undue tension, bearing and packing failures can be expected.

320

Repair And Maintenance

When dismantling a rotating cylinder, some important suggestions are:

1. Dismantle the cylinder on a clean workbench. Use proper tools to do the job. Do not use a pipe wrench on any of the finished surfaces. If the cylinder is placed in a vise, use soft pads and do not apply too much pressure on the cylinder body. Loosen the rod packing retainer, then remove the cover screws, take off the cover, and remove the piston and piston rod.

2. As the parts are removed, clean each part thoroughly. Any parts that have internal passages should be cleaned out

Courtesy Le Blond, Inc.

Fig. 32. Power chucking installation on a modern production lathe.

with compressed air. If the cylinder is to be dismantled for any length of time, use a protective coating on the steel or iron parts. If the parts are to be placed in storage, store them in a clean dry location.

3. If there are any score marks in the cylinder body, polish them out. If the scores are too deep, it may be necessary to scrap the body. If the body is made of cast iron, the score marks may be brazed and the interior of the body refinished.

4. If the piston rod is scored, polish out the scored marks with a fine emery cloth, unless they are too deep. If they are too deep, it is probably cheaper to replace the rod with a new one, since these rods are not long.

5. It is advisable to change all packings and seals when the cylinder is disassembled. Use those which are recommended by the manufacturer. If the piston uses cup packings, make certain that they are properly installed. Apply enough tension that they will not leak, but do not apply enough tension that the lips turn in. Some pistons are designed so that the piston parts make metal-to-metal contact, and too much cam pressure cannot be placed on the cups or they will be damaged.

6. If the rod bearing in the body is worn, it should be replaced. In some designs, the rod bearing is part of the body; then there is usually enough stock in the body to bore out and press in a bronze sleeve-type bearing. Keep in mind that the bearing must be tight.

7. If the air shaft or oil shaft is worn or scored, it should be replaced; this is a very vital part of the cylinder, and can be a great source of trouble if there are any rough parts on it.

8. In placing new packings in a cylinder, do not be afraid to use a little grease. This aids in installation, and provides a lubricant until the lubricators take over.

9. In reassembling the cylinder, make certain that all screws are tightened securely. Tighten the cover screws evenly. Be careful in placing the packing and gaskets, so that they are not cut.

10. After the cylinder is completely assembled, lubricate and turn the distributor body to make sure that it moves freely without any binding action, before pressure is applied. Make certain that the packings are not too tight.

11. Apply fluid pressure to one cylinder port of the distributor body, and check the piston movement. Check for leaks by placing a finger on the pipe port. Then shift the fluid pressure connection to the second port, let the piston move to the end of its travel, and place a finger on the first port to check for leaks. Any slight packing leaks that are caused by a slight blow of the shaft can usually be taken care of after the cylinder is run in, by pulling upward on the shaft packing. If excessive leakage occurs, the cylinder should be dismantled.

12. Mount the cylinder as instructed under installation procedure.

REVIEW QUESTIONS

1. What is the chief difference between a single-acting and a double-acting cylinder?

2. What is the chief difference between the ram-type cylinder and the piston-type cylinder?

3. Describe an application of a nonrotating cylinder.

4. Describe an application of a rotating cylinder.

5. What is a rod wiper? How is it installed? What is a rod scraper?

6. What is the purpose of a cushion collar and nose?

7. Explain how a cushion adjustment valve operates.

8. What are the advantages of a double-end cylinder?

9. What factors may cause a nonrotating cylinder to fail in service?

10. What factors may cause a rotating cylinder to fail?

11. List at least five types of rod packings that may be used in pneumatic or hydraulic cylinders.

12. Name at least five types of seals that are used on hydraulic cylinder pistons.

13. What type of piston seal is recommended for a hydraulic cylinder that must be as leakproof as possible? Why are these types of seals recommended?

14. Name at least three types of seals that can be used on pneumatic cylinder pistons.

15. Why should a boot be used to cover a piston rod of a pneumatic or hydraulic cylinder? How should the boot be fastened to the piston rod and cylinder cover?

16. What other protection may be provided the piston rod of a cylinder?

CHAPTER 15

Pneumatic Motors and Tools

Compressed air can be used for a wide variety of operations, such as: to drive air motors, to drive air hammers, to operate air cylinders, to atomize liquids, to convey materials, to displace liquids, to agitate liquids, to serve as a cushion, to create a vacuum, to accelerate combustion, to serve as a coolant, and to serve as a lubricant.

An electric motor is a device in which electrical energy is converted into mechanical energy. A *fluid motor* is a device for converting fluid energy into mechanical energy. In this chapter, the main features of pneumatic motors and pneumatic tools are discussed.

GENERAL TYPES OF AIR MOTORS

Air motors may be divided into two general types: (1) the rotary type and (2) the reciprocating type. The combination of a piston and cylinder is an example of the reciprocating type of motor. Two examples of the rotary motor are the vane-type and gear-type motors.

A rotary vane-type motor is illustrated in Fig. 1. The rotating member, with its sliding vanes, is set off center in the housing. As the air enters, it is trapped between the vanes (which ride on the inside of the housing) and pushes to the outlet. Thus the compressed air drives a rotating member.

A gear-type motor is illustrated in Fig. 2. A pair of meshed gears rotate inside a housing. As the gears rotate, the air is trapped between the gear teeth and housing, and pushes from the inlet to the outlet. Thus the compressed air drives the gears.

CHARACTERISTICS OF AIR MOTORS

An air motor operates safely in an explosive atmosphere, and has a low installation cost, a rapid acceleration, and a high starting torque. It is compact, and its variable speed can be changed.

A typical curve for torque versus speed (rpm) for a rotary-type air motor is illustrated in Fig. 3. Air-motor speed can be varied,

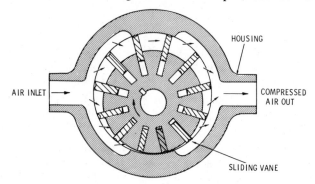

Fig. 1. Illustrating a sliding-vane type of air motor.

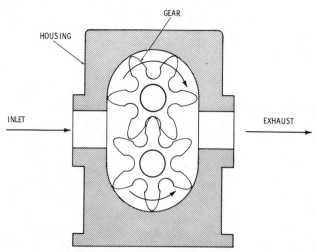

Fig. 2. Illustrating a gear-type air motor.

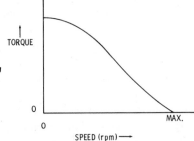

Fig. 3. Torque-speed curve for a rotary-type air motor.

in small steps, from zero to a maximum by varying the air flow to the motor. The air flow rate to the motor can be varied by means of a valve in the inlet compressed-air line. The pressure at the air motor inlet can be controlled with an adjustable pressure regulator in the air inlet line; thus torque can be controlled by means of the pressure regulator. An air motor can be arranged to be reversible in speed direction, by means of a four-way valve. As shown in Fig. 3, the speed decreases as the load increases. Air motors can be stalled without any damage being done.

A curve for horsepower versus speed for a rotary-type air motor is shown in Fig. 4. The power output is at a maximum between

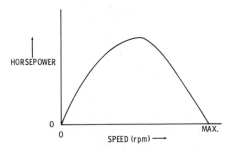

Fig. 4. Horsepower-speed curve for a rotary-type of air motor.

zero and maximum speed. Power output is proportional to the product of torque times angular or rotational speed (rpm).

The speed, torque, and direction of rotation of an air motor can be controlled and adjusted to job requirements. For example, reversible air motors are used to turn or screw studs into castings,

327

when assembling engines. The air pressure to the air motor is regulated, so that the air motor stalls when the stud is driven to the correct tightness. In this instance, the air motor starts off at high speed (very low torque), is quickly loaded so that it stalls, and is then reversed. An air motor has no tendency toward heating when it is overloaded; the motor can withstand repeated stalling and reversing without damage or overheating. An air motor can accelerate (or speed up) rapidly, because the compressed air energy can be released at a high rate.

Air motors are available commercially in sizes ranging from 1/10 horsepower to 50 horsepower, and in speeds ranging from 20 rpm to 75,000 rpm.

GENERAL FEATURES OF PNEUMATIC TOOLS

Pneumatic tools represent a large number of air-operated machines that are usually transported to a work area. Usually, the portable pneumatic tool is carried to the work, rather than transporting the work to the tool (as in machining some metals by machine tools). In many instances, the tool is either held entirely by means of the hands or guided and controlled directly by means of the hands. The tool may be located at a distance from the compressor. On an outdoor construction job, the compressor may be portable. In a manufacturing plant, the compressor may be stationary. Air from the compressor system is led by means of distribution piping through a filter, a pressure regulator, and a lubricator. Air from the lubricator may be led through a flexible hose to the pneumatic tool. Pneumatic tools are usually supplied with compressed air at about 90 pounds per square inch, gauge. There are three broad classifications of pneumatic tools: (1) rotary; (2) piston; and (3) percussion.

In a typical rotary pneumatic tool, the air enters the tool handle, passes through a manually operated control valve, passes through the end plates, and enters a chamber of a vane-type air motor. The air pushes the rotor from the inlet to the outlet. At the outlet, the air is discharged into the atmosphere. The typical rotary air motor possesses three to five blades, rotates at speeds of 4,000 to 25,000 rpm, and develops up to three horsepower. A spindle fastened to the rotor drives a set of planetary reduction gears to supply the desired output shaft speed. The output shaft may be equipped with a chuck for holding a drill, grinder, or some other tool.

The piston-type air motor is used for tools when higher horsepower and slower speeds are desired, as in hoists and heavy-duty drills.

The percussion- or hammer-type air motor is found in chippers, riveters, and pavement breakers. In this type of machine, the air usually passes through the handle, passes a control valve and into a cylinder with a reciprocating piston, passes through an exhaust valve, and then is exhausted into the atmosphere. The piston is moved back and forth. On the driving stroke, the inertia of the piston provides a hammer blow or impact to a drill, a chisel, or a bit.

Compressed air permits the use of tools which are relatively compact, light in weight, flexible, portable, and easy to operate. Pneumatic tools are used on very light assembly work and on heavy-duty construction jobs. The speed and direction of rotation of an air motor can be controlled quickly, and they can be adjusted to serve requirements by means of a control valve which throttles the flow of compressed air to the motor. Pneumatic motors do not have a tendency to become hot when they are overheated. An air motor presents no spark hazard in explosive atmospheres; therefore, it can be used in oil refineries and other plants which have the problem of explosive atmospheres. Air motors are used under wet and humid conditions, because there is no electric shock hazard.

Among the many applications and users of pneumatic tools are: agricultural equipment manufacturers, aircraft manufacturers, ammunition depots, automotive shops and plants, bakeries, boiler shops, can factories, cement plants, chemical plants, coal mines, construction work, cotton mills, cut-stone plants, dairies, distilleries, forge shops, foundries and steel mills, furniture factories, many kinds of industrial plants, laundries, machine shops, mining, oil refineries, paper mills, quarries, railroads, road building, rubber factories, sawmills, shipyards and boat-builders, smelters. structural work, and waterworks.

Courtesy Gardner-Denver Company

Fig. 5. Cutaway view of a pneumatic screwdriver.

PNEUMATIC TOOLS

From the point of view of application of pneumatic tools, two main classes of tools should be considered: (1) air-operated portable tools; and (2) air-operated rock drills.

Portable Tools

This class of pneumatic tools includes abrasive tools, drills, screwdrivers, hammers, riveters, and hoists.

Abrasive Tools—Abrasive tools include grinders, buffers, and sanders. A rotary vane-type air motor is usually used for tools of this type. Tools are available in a wide range of speed and power.

Drills—A rotary vane-type air motor is usually used for drills, reamers, tappers, and stud setters. Drills and reamers are used for steel, wood, and other materials. Some tools are reversible, with rotation possible in either direction, and some tools are nonreversible. Drill speed can be varied by throttling the air supply to the drill. Portable drills powered by a vane-type rotary air motor are used also for drilling any mineral substance that can be penetrated by an auger or a tungsten carbide bit—for example, coal, chalk, shale, and hardpan.

Screwdrivers—A rotary vane-type air motor is usually used for screwdrivers, nut setters, impact wrenches, shears, and nibblers (Fig. 5). Millions of nuts, bolts, and screws are driven daily by pneumatic tools in the large production plants making products, such as automobiles, refrigerators, radios, and various appliances. Pneumatic impact wrenches are used to remove or to tighten nuts by torsional or rotary impacts, and pneumatic nibblers and shears are used for cutting and shearing steel metal.

Hammers—In the pneumatic scaling or chipping hammer, a piston delivers a series of blows to a forming tool or chisel at the end of the hammer; the piston is air-operated. The two types of percussion tools are the valve type and the valveless type. In the valve type, a valve arrangement is used to control the flow of air both to and from the two ends of the hammering piston. In the valveless type, the valve action is performed by the piston. The valve-type of hammer is used for chipping and riveting, and the valveless type is used for removing scale. Sand rammers are actually pneumatic hammers; they are used in foundries for packing sand molds.

Hoists—Hoists that are operated by compressed air are used in many applications, especially in machine shops and in foundries. Pneumatic hoists are used outdoors and in conditions where fumes and explosive gases are present. An air motor is variable

331

in speed, reversible in direction, and can withstand stalling from an overload without damage.

Rock Drills and Paving Breakers

The second class of air-operated tools includes various types of rock drills and pavement breakers. The hammer drill is a commonly used air-operated tool in mining and general excavation work. A piston can be found inside the air cylinder, and a drill

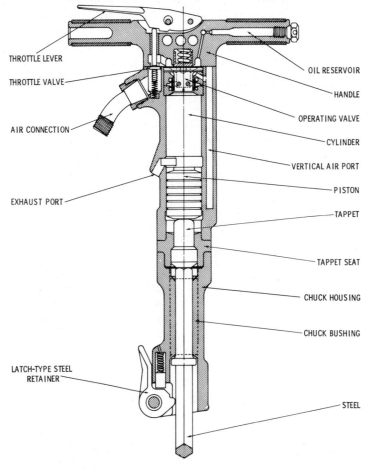

THROTTLE LEVER

THROTTLE VALVE

AIR CONNECTION

EXHAUST PORT

LATCH-TYPE STEEL
RETAINER

OIL RESERVOIR

HANDLE

OPERATING VALVE

CYLINDER

VERTICAL AIR PORT

PISTON

TAPPET

TAPPET SEAT

CHUCK HOUSING

CHUCK BUSHING

STEEL

Courtesy Compressed Air and Gas Institute

Fig. 6. Sectional view of a paving breaker.

steel is used. The drill steel is provided with a shank on one end and a bit at the other end. The freely moving piston hammers or strikes rapid blows on the shank of the drill steel.

Paving breakers are hand-operated pneumatic tools that have a wide variety of uses in industrial plants and in general construction work. One type of tool is valveless; another type of tool is valve operated (Fig. 6).

One class of pneumatic tools is sometimes designated as pneumatic specialty tools. The pneumatic specialty tools include: staybolt riveters, paint mixers, screw spike drives, wire-wrapping devices, spinner riveters, shank grinders, reciprocating filing machines, valve grinders, staybolt tappers, air cranking motors, cut-off machines, concrete vibrators, countersinking tools, fender irons, hog ringers, clip squeezers, tire-valve inserters, air-powered generators, multiple-nut runners, multiple screwdrivers, core-box vibrators, and core-knockout machines.

REVIEW QUESTIONS

1. In some applications, as in mining, only pneumatic tools are used. List several reasons for this fact.

2. Pneumatic drills are used in many light assembly jobs which employ women. Explain why a pneumatic drill is preferred to an electric drill for these jobs.

3. Explain several advantages of portable air tools for street construction and repair work.

CHAPTER 16

Rotary Hydraulic Motors and Hydraulic Transmissions

In a fluid pump, mechanical energy is converted to fluid pressure energy. In a fluid motor, fluid pressure energy is converted to mechanical energy. In this chapter, fluid motors and the combination of a fluid pump and a fluid motor are discussed.

GENERAL TYPES OF FLUID MOTORS

The rotary fluid motor is used in many applications. The fluid motor may be a gear-type, vane-type, or piston-type motor. These motors are similar in construction to gear-type, vane-type, and piston-type pumps. Rotary motors may be either the fixed-displacement or the variable-displacement type.

Gear-Type Motor

The principle of operation of a gear-type motor is illustrated in Fig. 1. The operation is practically the same as the reverse action of a gear-type pump. Hydraulic oil, at high pressure, enters at the inlet, pushes each of the gears, and then flows outward. Usually, the load that is to be moved is connected to only one of the gears. A gear-type motor (Fig. 2) is a fixed-displacement machine. Gear-type motors are commercially available for pressures up to 3000 pounds per square inch and for horsepower outputs up to 200 horsepower. In most instances, the gears are spur gears; however, in some instances, the gears may be helical or herringbone gears.

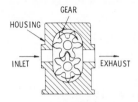

Fig. 1. Principle of operation of a gear-type hydraulic motor.

Vane-Type Motors

The basic action of a rotary vane-type motor is illustrated in Fig. 3. Oil, at high pressure, enters at the inlet, pushes on the vanes to rotate the rotor, and passes through to the outlet. Some method must be devised for holding the vanes against the contour of the housing at the start. This can be done either by fluid pressure or by means of springs. In a vane-type pump, centrifugal force holds the vanes against the housing. The vane-type motor (Fig. 4) is a fixed-displacement machine. Vane-type motors are commercially available for pressure as high as 2000 pounds per square inch and for output horsepower as high as 300 horsepower, and more.

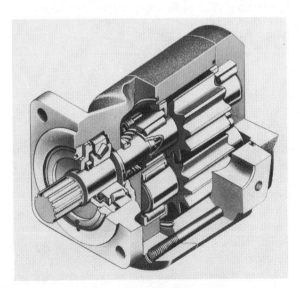

Courtesy Commercial Shearing, Inc.

Fig. 2. Sectional view of a gear-type pump or motor for 2,000 psi.

336

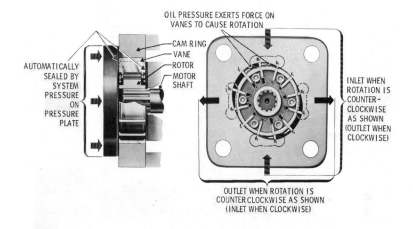

OIL PRESSURE EXERTS FORCE ON
VANES TO CAUSE ROTATION

CAM RING
VANE
ROTOR
MOTOR
SHAFT

AUTOMATICALLY
SEALED BY
SYSTEM
PRESSURE
ON
PRESSURE
PLATE

INLET WHEN
ROTATION IS
COUNTER-
CLOCKWISE
AS SHOWN
(OUTLET WHEN
CLOCKWISE)

OUTLET WHEN ROTATION IS
COUNTER CLOCKWISE AS SHOWN
(INLET WHEN CLOCKWISE)

Courtesy Sperry Vickers, Division of Sperry Rand Corporation

Fig. 3. Operating principle of a rotary vane-type motor.

Piston-Type Motors

Piston-type motors are available in both radial and axial designs. A radial piston-type constant-displacement hydraulic motor

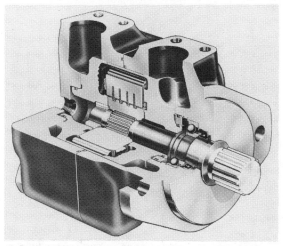

Courtesy Sperry Vickers, Division of Sperry Rand Corporation

Fig. 4. Sectional view of a rotary vane-type motor.

337

is illustrated in Fig. 5. The cylinder rotates around the fixed pintle. Oil, at high pressure, enters the upper ports in the pintle; this forces the pistons to move outward, causing the cylinder and output shaft to rotate in a clockwise direction.

The basic operation of an axial piston-type motor is diagrammed in Fig. 6. Oil, at high pressure, enters the stationary valve plate. This oil forces the pistons outward, thus rotating the output shaft.

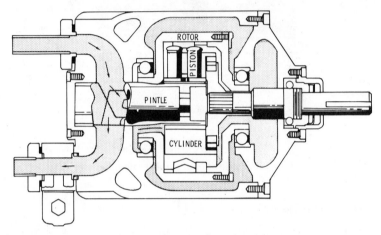

Courtesy The Oilgear Company

Fig. 5. Radial piston-type motor.

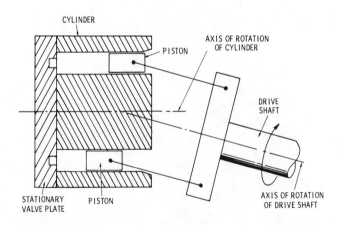

Fig. 6. Basic principles of operation of an axial piston-type motor.

The axial piston-type motor can be either a fixed-displacement or a variable-displacement motor, depending on the angle between the cylinder axis and the output shaft. If this angle can be varied, the motor is a variable-displacement motor.

A variable-displacement axial piston-type motor is shown in Fig. 7. Oil, at high pressure, enters the motor at the inlet port.

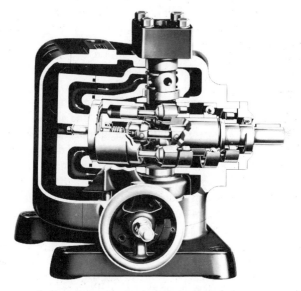

Courtesy Sperry Vickers, Division of Sperry Rand Corporation

Fig. 7. Variable-displacement axial piston-type motor.

passes through the pintle, yoke, valve block, and inlet port of the valve plate, and then passes into the cylinder. Oil, at high pressure, pushes the pistons away from the valve plate, causing the cylinder and output shaft to rotate. The oil leaves through the outlet port of the valve plate, passing through the yoke, outlet pintle, and outward through the discharge flange. The angle between the cylinder axis and the output shaft can be varied from 7 1/2° to 30°; therefore, the displacement can be varied from a minimum quantity to four times the minimum quantity.

Piston-type motors are available commercially for oil pressures as high as 5000 pounds, and more, per square inch. Horsepower outputs range to 150 horsepower, and more.

339

FLUID TRANSMISSIONS

Various methods are employed for coupling or joining two rotating shafts—for example, connecting an electric motor or an automotive engine to a load. Strictly mechanical couplings that are either rigid or flexible are used in various applications. In some applications, however, performance and service requirements are best satisfied by some form of fluid connection.

Hydrodynamic

A so-called "hydrodynamic" transmission depends on fluid dynamic or velocity changes, rather than on a displacement action. A fluid coupling is used in one type of hydrodynamic transmission; basically, this is a combination of a centrifugal pump and a reaction turbine. Oil is usually employed as the fluid in commercial units, because it possesses lubricating properties, stability, and availability. The basic principle of this type of hydrodynamic fluid coupling can be illustrated by means of two ordinary electric fans placed to face each other. One of the fans, which is connected to an electric outlet, is set in motion by turning on the electric current. As its blades are rotated, the air flow that is developed rotates the blades of the second fan, which receives no electric current.

Hydrostatic

A hydrostatic transmission, or hydraulic displacement-type transmission, is a combination of two interconnected positive-displacement units—a pump and a motor. This is a contrast to the hydrodynamic type of transmission.

The pump may be located at a considerable distance from the motor; or the two units (pump and motor) may be combined in a single housing. The rotary hydraulic pump transforms mechanical energy to fluid pressure energy. The high-pressure oil is then delivered to a rotary hydraulic motor. In the motor, the fluid pressure energy is transformed to mechanical energy. The oil flows from the pump to the motor; it then flows from the motor to the pump.

This type of transmission has several advantages: (1) output shaft speed can be maintained at a constant speed for a variable-speed input; (2) speed and direction of the output shaft can be controlled accurately and remotely; (3) constant power output can be maintained over a wide range of speeds; (4) automatic torque control at the output shaft can be maintained; (5) output shaft speed can be varied in extremely small steps; (6) the output shaft can be reversed quickly and without shock; (7) power consumption can be kept at a low level; and (8) automatic overload protection can be maintained.

As illustrated in Fig. 8, a variable-displacement pump can be connected to a fixed-displacement motor. If the work load is constant, a variation in the output shaft speed results in an almost constant output torque and in variable power. Constant-torque transmissions are used on machine tool feeds and conveyors.

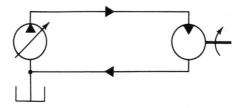

Fig. 8. Schematic diagram of variable-displacement pump connected to fixed-displacement motor.

A fixed-displacement pump can be connected to a fixed-displacement motor (Fig. 9). If a flow control is placed in the line to vary the rate of oil flow to the motor and if the load on the output

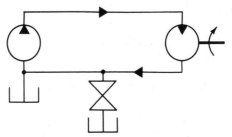

Fig. 9. Schematic diagram of fixed-displacement pump and fixed-displacement motor.

shaft is constant, the torque is constant and the power can be varied with the speed.

A fixed-displacement pump can also be connected to a variable-displacement motor (Fig. 10). If it is desired to maintain constant

341

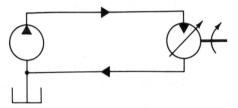

Fig. 10. Schematic diagram of fixed-displacement pump and variable-displacement motor.

power output as the output speed is varied—as in winders and machine tool spindle drives, for example, the displacement of the motor can be varied; then the output torque decreases as the speed increases.

A variable-displacement pump connected to a variable-displacement motor is shown in Fig. 11. This combination can provide variations in output shaft speed, torque, and power characteristics of hydraulic transmissions.

A brief summary of the characteristics of hydraulic transmissions is shown in Table 1.

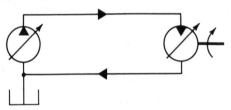

Fig. 11. Schematic diagram of variable-displacement pump and variable-displacement motor.

Table 1. Summary Of Characteristics Of Hydraulic Transmissions

Displacement		Output		
PUMP	MOTOR	TORQUE	SPEED	POWER
Fixed	Fixed	Constant	Constant	Constant
Variable	Fixed	Constant	Variable	Variable
Fixed	Variable	Variable	Variable	Constant
Variable	Variable	Variable	Variable	Variable

The displacement-type transmission (Fig. 12) includes a variable-displacement pump and a fixed-displacement motor. Both the hydraulic pump and the hydraulic motor are radial piston-type devices.

The input shaft is driven from any constant-speed source of power. This rotation is transmitted through a splined floating coupling flange directly to the cylinder barrel mounted on the

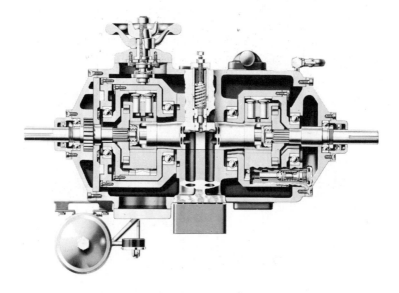

Courtesy The Oilgear Company

Fig. 12. Displacement-type transmission.

fixed pintle. The radial pistons in the driven cylinder are confined within the rotor by means of concave reaction rings; the rotor revolves on antifriction bearings in the adjustable-stroke slide block.

When the center lines of the cylinder and the rotor coincide, no reciprocating motion is imparted to the piston; thus no oil is delivered from the pump. When the center line of the cylinder does not coincide with that of the rotor, the pump delivers oil to the motor. The handwheel is used to control the variable delivery of the pump. When oil is delivered to the motor, the output shaft is rotated.

Hydrodifferential Transmissions

Various methods have been devised to improve the efficiency of hydrostatic transmissions. The hydraulic differential transmission is a general arrangement in which a portion of the energy is transmitted mechanically. There are two types of hydrodifferential transmissions: (1) the split-torque (or input-coupled) type; and (2) the split-speed (or output-coupled) type.

Split-Torque—A split-torque hydrodifferential transmission is illustrated in Fig. 13. A torque T_1 is applied at the input shaft

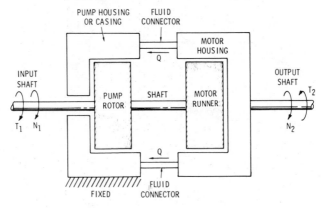

Fig. 13. Split-torque hydrodifferential transmission.

which rotates at the input speed N_1. A portion of the torque drives the motor runner. The pump housing or casing is fixed in position. The pumping action is proportional to the relative motion between the rotor and the housing. Fluid passes from the pump housing to the motor housing, and then returns to the pump at the volume rate of flow Q. The motor housing rotates the output shaft at the output shaft speed N_2 with the output shaft torque T_2. For ideal performance with no losses, the pump displacement D_1 (in cubic inches) and the motor displacement D_2 (in cubic inches) are related to speed and torque, as shown in the formulas:

$$\frac{N_2}{N_1} = 1 + \frac{D_2}{D_1}$$

and;

$$\frac{T_2}{T_1} = \frac{1}{1 + \dfrac{D_1}{D_2}}$$

Split-Speed—A split-speed hydrodifferential transmission is diagrammed in Fig. 14. The input shaft with the torque T_1 applied

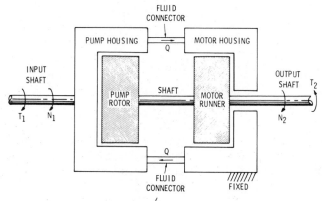

Fig. 14. Split-speed hydrodifferential transmission.

to rotate the input shaft at the input speed N_1 drives the housing of the pump. The pump rotor drives the motor runner; the output shaft torque is T_2 and the output shaft rotates at the output shaft speed N_2. The motor housing is fixed in position. Fluid

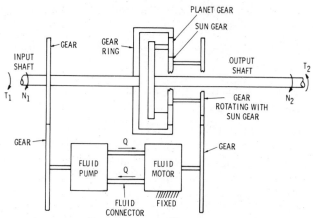

Fig. 15. Planetary hydrodifferential transmission.

passes through the fluid connection from the pump housing to the motor housing, and returns to the pump housing. For ideal performance with no losses, the pump displacement D_1 (in cubic inches) and the motor displacement (in cubic inches) are related to speed and torque, as shown in the formula:

$$\frac{N_2}{N_1} = \frac{1}{1 + \dfrac{D_2}{D_1}}$$

and;

$$\frac{T_2}{T_1} = 1 + \frac{D_2}{D_1}$$

A hydrodifferential transmission with a planetary gear train is diagrammed in Fig. 15. The system consists of a variable positive-displacement pump and a positive-displacement motor, interconnected with a planetary gear train.

REVIEW QUESTIONS

1. Displacement-type hydraulic transmissions are sometimes used to move gun turrets. What are the advantages of this type of transmission for this application?

2. Is a displacement-type hydraulic transmission suitable for an automobile? Explain.

3. How do hydraulic motors compare in size with electric motors for the same horsepower output?

4. Name four applications for fluid motors, and explain how they are used.

5. Make a rough sketch, using *ANS* Graphic Symbols, of a simple circuit which incorporates the following: hydraulic power device with constant-displacement pump; a four-way two-position manually operated valve; and a fixed-displacement type of hydraulic motor.

6. Explain how a foot-mounted type of hydraulic motor should be mounted.

CHAPTER 17

Accumulators

An *accumulator* is found in many hydraulic systems. The accumulator, as the name suggests, is a storage device. The various types of accumulators are studied in this chapter.

A simple accumulator is sometimes used in household water systems (Fig. 1). The accumulator may consist of a tee with a side branch pipe that is capped. The air that is trapped in the side branch pipe is compressed, and then acts like a compressed spring.

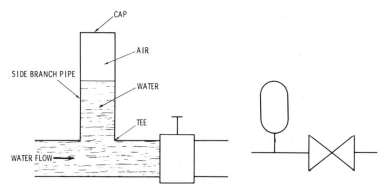

Fig. 1. Diagram showing the use of a simple accumulator in a household water system.

As a faucet is either opened or closed quickly, a sudden change in pressure and flow occurs. The trapped air acts as a cushion, or shock absorber, to prevent water hammering in the piping system.

The storage battery in a car is a typical example of an electrical or chemical accumulator. Chemical energy is stored in the battery when the battery is not in use. The stored chemical energy is converted into electricity that is used to start the engine.

HYDRAULIC ACCUMULATORS

A hydraulic accumulator may be used for a variety of purposes. Some of its uses are: (1) as a shock absorber; (2) to provide oil makeup in a closed system; (3) to compensate for leakage in a system; (4) to provide a source of emergency power in event of failure of the normal power supply; (5) to maintain steady delivery pressure over a period of time without keeping the pump operating continuously; and (6) as a transfer barrier device to separate the oil from some other fluid in the system.

TYPES OF ACCUMULATORS

Accumulators may be divided into three general types: (1) weight-loaded; (2) spring-loaded; and (3) air- or gas-type accumulators. The air- or gas-type accumulator can be subdivided further into the separator and nonseparator types.

Weight-Loaded or Gravity-Type

The weight-loaded or gravity-type accumulator (Fig. 2) consists of a cylinder, a movable piston, a ram or plunger, and a weight.

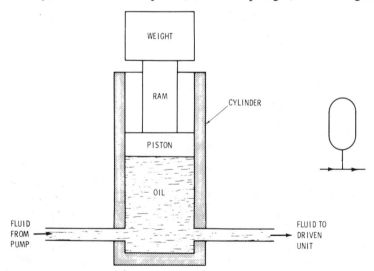

Fig. 2. Diagram of a weight-loaded or gravity-type accumulator.

348

The dead weight (which may be placed in a container) may be concrete, iron, steel, water, or other heavy material. The piston should have a precision fit inside the cylinder in order to reduce leakage. The inner cylinder wall should have a honed or ground finish in order to reduce friction and wear. As hydraulic oil is pumped into the cylinder, the piston pushes the weight to a higher level. Thus the potential or stored energy of the weight is increased. The energy stored in the weight is released in the downward motion as it is required by the demands of the system. An accumulator of this type may be custom-built for a particular installation. The weight is adjusted so that the ram rises when the fluid pressure reaches a set level. The travel of the ram can be controlled by an arrangement of a cam on the plunger and limit switches. The gravity force of the piston on the oil provides a nearly constant oil pressure level for the full stroke of the piston. By providing adequate piston area and ample length of piston stroke, a large volume of fluid can be supplied at high pressure. A single large accumulator may provide service for a number of different machines.

Spring-Loaded Type

A spring-loaded type of accumulator is illustrated in Fig. 3. This device consists of a cylinder, a piston, and a spring. One or

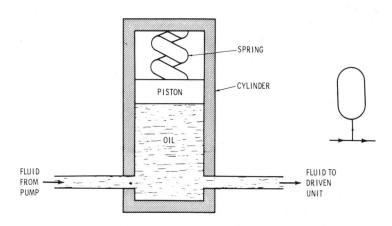

Fig. 3. Diagram of a spring-loaded type of accumulator.

more springs may be used. The springs may be arranged to provide various adjustments by means of bolts. As the oil is pumped into the accumulator, the piston or plunger compresses the spring; thus energy is stored in the spring. The energy stored in the spring is released as it is required by the demands of the system. The pressure on the oil is not constant for all the positions of the piston, because the spring force depends on the movement of the spring. Usually, this type of accumulator delivers only a small amount of oil at low pressure.

Air or Gas Type

Hydraulic fluid or oil is nearly incompressible. This means that a large increase in oil pressure results in only a small, or negligible, decrease in the volume of oil. On the other hand, a large increase in air or gas pressure results in a large decrease in the volume of the air or gas. Relatively speaking, hydraulic oil is less elastic or spring-like than air. Oil cannot be used effectively to store energy by compressing it, whereas air or gas can be compressed to store energy. Thus, one general type of accumulator uses a gas or air, rather than a mechanical spring or a weight, to provide the spring-like action. Air or gas types of accumulators can be divided into two subdivisions: (1) the nonseparator type; and (2) the separator type. In the nonseparator type of accumulator, the oil is in direct contact with the air or gas. In the separator type of accumulator, some type of mechanical material or device is used to separate the air or gas from the oil. In the separator type of accumulator, either a solid or a flexible barrier is placed between the oil and the air or gas to separate the two different types of fluids.

Nonseparator Type—A nonseparator type of air or gas accumulator (Fig. 4) consists of a fully enclosed cylinder, adequate ports, and a charging valve. A portion of the oil must be trapped in the bottom of the cylinder before this type of accumulator can be placed in operation. Air, nitrogen, or an inert gas is then forced into the cylinder, and the accumulator precharged to the minimum pressure requirements of the system. A so-called "free surface" exists between the oil and the air or gas. As a greater quantity of oil is pumped into the accumulator, the air or gas above the oil is compressed still further. The energy is stored in the compressed gas, and it is released as required by the demands of the system.

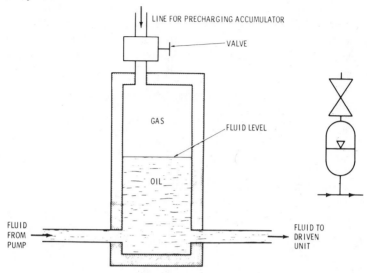

LINE FOR PRECHARGING ACCUMULATOR

VALVE

GAS

FLUID LEVEL

OIL

FLUID
FROM
PUMP

FLUID TO
DRIVEN
UNIT

Fig. 4. Illustrating a nonseparator type of air or gas accumulator.

This type of accumulator should be mounted in a vertical position, because the gas must be retained in the top of the cylinder. To prevent the air or gas being exhausted into the hydraulic system, only about two-thirds of the accumulator volume can be used for the air or gas volume. Approximately one-third of the remaining accumulator volume should be reserved for the oil, to prevent the air or gas from being drawn out of the accumulator into the hydraulic system. Aeration, or mixing, of the oil and air or gas may result in diminishing the precharge of the accumulator. If the air or gas is absorbed by the oil, the accumulator will not function properly. The nonseparator type of accumulator requires an air or gas compressor for the precharging operation of the accumulator.

Separator Type (With Piston)—A separator type of accumulator with a free or floating piston acting as the barrier between the air or gas and the oil is illustrated in Fig. 5. High-pressure air or gas is charged into the space on one side of the piston, and hydraulic oil is charged into the space on the opposite side.

The tube should be machined with precision. The piston packing keeps the oil and gas separated.

This type of accumulator may be installed in any position. The preferred position, however, is to place the cylinder axis vertically, with the gas connection at the top. The wearing action of the packing between the piston and the cylinder should be checked after extended use, because this may result in significant leakage.

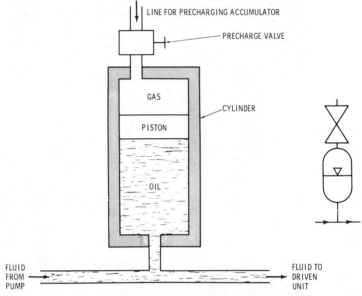

Fig. 5. Illustrating a piston-type accumulator.

A floating piston within a cylindrical accumulator is illustrated in Fig. 6. In this design, the double-shell construction provides a pressure-balanced inner shell which contains the piston and serves as a separator between the precharge air or gas and the working hydraulic fluid. The outer shell serves as a gas container. Rapid decompression of the precharged air or gas, resulting from a rapid discharge of the working hydraulic fluid, provides a coolant for the entire working area of the inner shell. Pressure-balancing ports in the piston provide equal pressure to either side of both rings; this prevents pressure lock between seals. The types of information which a manufacturer can supply to help the user are shown in Fig. 6.

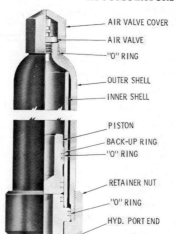

Fig. 6. A piston-type accumulator.

AIR VALVE COVER

AIR VALVE

"O" RING

OUTER SHELL

INNER SHELL

PISTON

BACK-UP RING

"O" RING

RETAINER NUT

"O" RING

HYD. PORT END

Courtesy Superior Hydraulics, Cleveland, Ohio

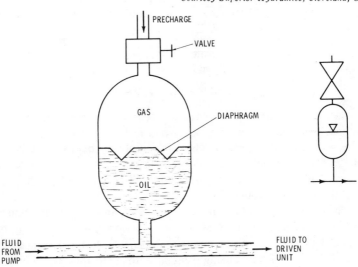

PRECHARGE

VALVE

GAS

DIAPHRAGM

OIL

FLUID
FROM
PUMP

FLUID TO
DRIVEN
UNIT

Fig. 7. A diaphragm-type accumulator.

Separator Type (With Diaphragm)—A diaphragm-type accumulator (Fig. 7) involves two hemispheres that are made from steel forgings. The hemispheres are locked together, and a flexible, convoluted, rubber diaphragm is clamped around the periphery. An

353

air or gas precharge is applied to one of the two hemispheres; oil, under pump pressure, is applied to the opposite hemisphere to compress the air or gas charge. As the air or gas is compressed, the pressure rises; then the gas acts as a spring. Oil pressure and gas pressure are equal, because the separating member is flexible.

Bag Type—The bag, or bladder, type of accumulator is a seamless steel shell that is cylindrical in shape and spherical at both ends. A gas valve is located at one end of the shell and opens into the shell. A large opening through which the bag can be inserted is located at the opposite end. The bladder is made of synthetic rubber and is pear shaped. The fully enclosed bladder, including a molded air stem, is fastened by means of a lock nut to the upper end of the shell. On the opposite end of the shell, a plug assembly containing the oil port and a poppet valve is mounted. The accumulator cannot be disassembled while a gas charge is inside the bag. The accumulator should be installed with the end that contains the air at the top to avoid trapping the oil when discharging.

CONTAMINATION

As with the other components in a hydraulic system, care should be taken to avoid contamination when installing the accumulator. The accumulator should be cleaned completely before installation in the system. Since an accumulator usually forms a dead-end in the pipe line, it may not be flushed as well as some other components during system operation.

The piston-type accumulator (see Figs. 5 and 6) can be studied to better understand accumulator construction. The piston may be sealed with an "O" ring. Contaminations may become trapped in the "O" ring groove, causing wear or damage to the "O" ring and allowing the gas charge to leak into the oil. An arrangement in which an "O" ring seal is placed between *Teflon* piston rings provides a means by which the piston rings can scrape contaminants from the cylinder walls, and thus protect the "O" ring seal. It is a good practice to install a filter in the air or gas charging system of the accumulator.

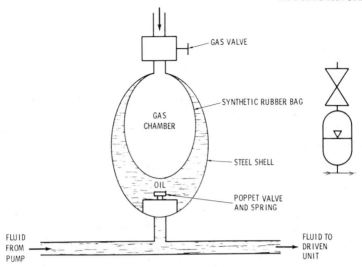

Fig. 8. A bag or bladder type of accumulator.

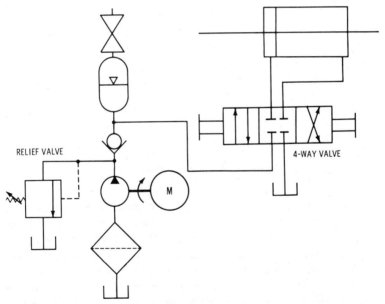

Fig. 9. Schematic diagram of a system in which the accumulator is used to absorb shocks.

ACCUMULATORS IN SYSTEMS

The accumulator is often installed in a hydraulic system to absorb shocks. As shown in the schematic diagram in Fig. 9, the hydraulic oil from a pump is piped to a four-way valve, which directs the oil flow to a cylinder containing a piston. If the valve is closed quickly, the sudden stoppage may result in oil shock waves—or a hammering effect. A violent hammering action may damage the fittings and piping. The accumulator is capable of absorbing the shocks, thus protecting the entire system.

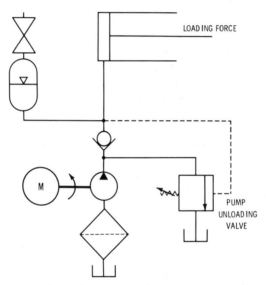

Fig. 10. Schematic diagram of a system that includes a power-saving accumulator.

During periods when no flow of oil is required in the hydraulic system, pump delivery can be returned, or by-passed, to the reservoir at low pressure. This arrangement serves to reduce electric power requirements, oil heating, and wear of the moving parts.

The accumulator can be used as a power-saving device, as illustrated in the schematic diagram in Fig. 10. A piston provides a loading force—as for a clamping operation or a rolling mill operation, in which the force moves only a short distance. After the oil pressure is built up at the piston face, the accumulator can supply the loading force for a period of time. During this period it is unnecessary to keep the pump delivering at high pressure. Therefore, an unloading valve is provided to return the pump delivery to the reservoir at low outlet pressure. During the unloading process, the pump discharge pressure is at a low level, while the accumulator pressure is at a high level to provide the loading force. Thus, the accumulator is a power-saving device, and also provides for absorption of shock waves.

REVIEW QUESTIONS

1. What is an accumulator?

2. Describe a gravity-type accumulator.

3. What is the chief difference between a gravity-type accumulator and a spring-type accumulator?

4. What are the advantages of an air or gas type of accumulator?

5. How does a piston-type accumulator differ from a hydraulic cylinder?

6. What is the chief advantage of using an accumulator in the hydraulic system of a curing press, where pressure must be held for long periods of time?

7. Explain how an accumulator can be used as an emergency source of power in a hydraulic system.

8. What type of accumulator is recommended for a central hydraulic system? Why?

CHAPTER 18

Pneumatic Circuits

In studying either pneumatic or hydraulic fluid power circuits, it is recommended that the less complicated circuits be studied before advancing to the more complicated circuits. This chapter discusses the basic circuits as well as the more advanced pneumatic circuits. A problem that is to be solved should be analyzed before proceeding to solve it by the simplest possible method. Overloading a circuit with unnecessary components is an unwise practice, and should be avoided.

In the schematic diagram (Fig. 1), the problem is to move an object A from position X to position Y at a nearly constant speed. To perform this operation, the operator shifts the handle of the two-position four-way valve C to direct air from the inlet port of the valve to port 1, which provides air to port 3 of cylinder D. This forces the piston of the cylinder to move forward. As the air pressure forces the piston forward, air is exhausted from port 4 of the cylinder to port 2 of the valve, passing through the exhaust port into the atmosphere. The movement of the piston and piston rod of the cylinder advances the object from position X to position Y.

To retract the piston, the operator shifts the handle to the second position of the valve (see Fig. 1); this directs the air from the inlet of the valve to port 2 and then to port 4 of cylinder D. As the piston of the cylinder begins to retract, air is exhausted through port 3 to port 1 of valve C, and then passes through the exhaust port of the valve into the atmosphere. The piston rod of the cylinder retracts fully to position X to complete the cycle. If the noise created by the air exhausting from the exhaust

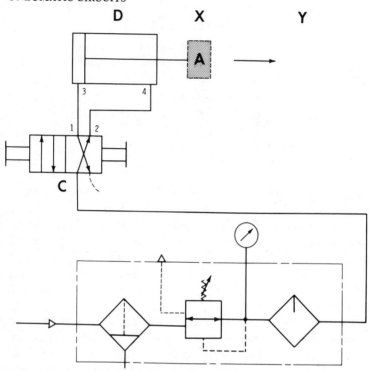

Fig. 1. Schematic diagram of a circuit for moving an object at a nearly constant speed.

port of the valve is objectionable, a muffler may be connected to the exhaust port. The capacity of the muffler should be large enough to prevent back pressure inside the valve.

Although the valve is actuated manually, in this instance, a solenoid or pilot actuator may be used. Many types of actuators are available, depending on the requirement of the user.

If it is necessary to control the rate of speed for moving the object A from position X to position Y, a flow control E can be added (Fig. 2). The flow control controls the flow of air exhausting from port 4 of the cylinder, which, in turn, controls the speed of the piston of cylinder D. On the return stroke of the piston, the check in the flow control opens, and the piston returns at full speed. The orifice in the flow control can be adjusted by means of a needle to permit different speeds.

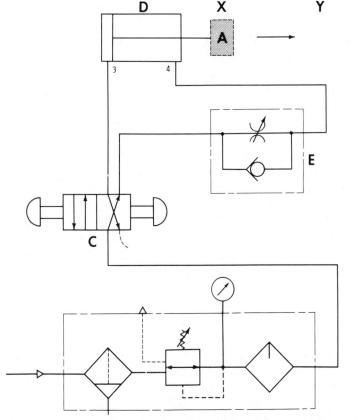

Fig. 2. Controlling the rate of speed of the piston by adding a flow control to the circuit.

If it is necessary to begin the movement of the object *A*, from position *X* at a given speed and then approach the position *Y* at a reduced speed (Fig. 3), either a cam-actuated or a solenoid-actuated speed control *F* can be installed in the system. When the trip contacts the actuator on the speed control, the speed changes automatically. Since the speed control contains a built-in check valve, the piston in the cylinder can return at full speed. Several tripping mechanisms can be used in conjunction with the piston rod, or other moving member connected to the piston rod, to provide a skip feed which is used in many machining operations.

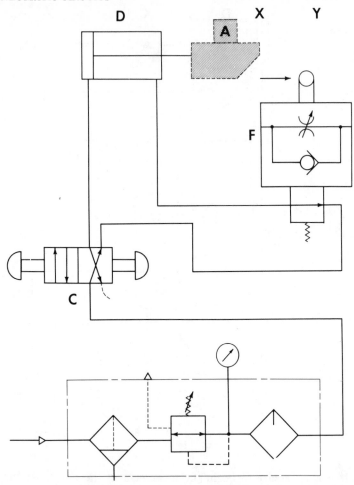

Fig. 3. Schematic diagram of a circuit in which the piston is moved at a given speed, and then the speed is reduced.

In some applications, the operator may desire to operate only one actuator for the piston in the cylinder to move through a complete cycle (Fig. 4). A complete cycle is the complete advancement and retraction of the piston and piston rod in the cylinder.

As shown in Fig. 4, a pilot-operated four-way valve C and two three-way valves B and D can be installed in the circuit, so

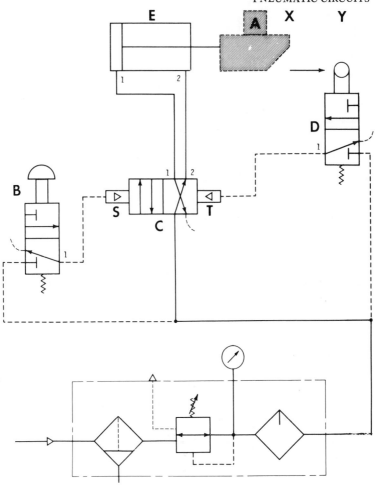

Fig. 4. Schematic diagram of a circuit in which the operator is required to actuate only a single valve to accomplish a complete cycle.

that the operator can actuate a single valve B to accomplish the complete cycle. The operator can momentarily shift the handle of the manually controlled three-way valve B to direct air from the inlet port to the cylinder port 1 and then to the pilot connection S of the four-way valve C. The spool in the valve shifts to direct pressure from the inlet port to cylinder port 1 and then to port

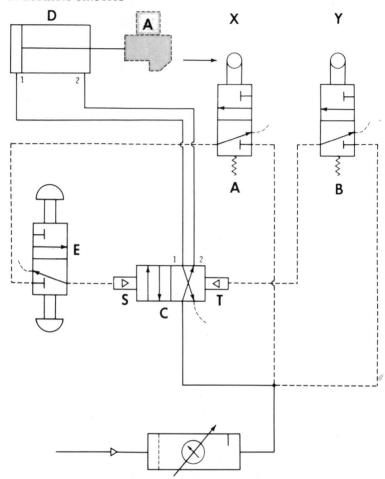

Fig. 5. Schematic diagram of a circuit in which the piston rod in the cylinder is required to reciprocate a number of times to produce a number of cycles.

1 of cylinder E. The piston advances, and air is exhausted from port 2 of the cylinder to port 2 of the four-way valve C and then onward to the exhaust port and into the atmosphere. When the piston in cylinder E advances to the end of the stroke, the cam roller on the cam-actuated three-way valve D is contacted, and air is directed from the inlet port to port 1 and onward to the

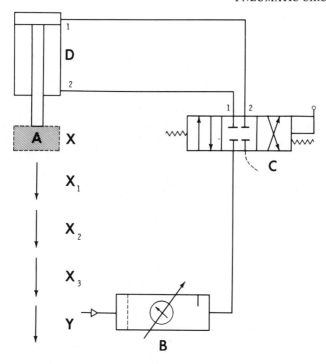

Fig. 6. Schematic diagram of a circuit designed to stop the piston at intermediate points between the extreme ends of the stroke.

pilot connection T of the four-way valve. The spool in the four-way valve shifts to its original position, and air is directed from the inlet port of the valve through cylinder port 2 and onward to port 2 of cylinder E. The piston retracts in the cylinder as air is exhausted from port 1 of the cylinder to port 1 of valve C and through the exhaust port to the atmosphere to complete the cycle.

Other types of valves can be substituted in the circuit shown in Fig. 4. Two electrical switches (one manually operated and the other cam-operated) and a solenoid-operated four-way valve can be installed in the circuit to accomplish the same operation.

A circuit can also be devised to cause the piston rod in the cylinder to reciprocate a number of times to produce a number of cycles (Fig. 5). When the operator shifts the handle of the manually operated three-way valve E, air flows to the pilot connection S of the four-way valve C. The spool of the valve shifts and air from the valve inlet is directed to the cylinder port 1 of the valve and onward to port 1 of the cylinder D. The piston of the cylinder advances as air is exhausted from port 2 of the cylinder to port 2 of valve C and to the atmosphere through the exhaust port. As the trip mechanism on the end of the piston rod reaches position Y, the cam roller on the cam-actuated three-way valve B is depressed, and air is directed from the inlet to the cylinder port of the valve and onward to the pilot connection T of valve C. The spool of valve C shifts to its original position, and air is directed from the inlet port to the cylinder port 2 and then to port 2 of the cylinder D. The piston retracts as air is exhausted from port 1 of the cylinder to port 1 of the valve, and passes outward through the exhaust port. When the piston is fully retracted to the point X where the trip mechanism on the piston rod contacts the roller of valve A, air is directed from the inlet port of the valve A through the cylinder port to the manually operated valve E and onward to the pilot connection S of valve C; the spool of valve C shifts and begins the next cycle. The reciprocating action of the piston is continuous, until the operator shifts the handle of the valve E, which causes the piston to stop in the retracted position. This type of circuit can be used on a saw, a polisher, a mixer, or other applications that require automatic reciprocation.

In applications where it is necessary to stop the piston at intermediate points between positions X and Y, the circuit shown in Fig. 6 can be employed. It is rather difficult to stop an air piston at any point other than the extreme ends of the stroke; however, it is relatively simple to stop either an air-hydraulic piston or a hydraulic piston within fairly close limits, because a hydraulic fluid is noncompressible for most purposes.

When the operator shifts the handle of the spring-centered, three-position, four-way valve C (see Fig. 6), air is directed from the inlet to cylinder port 2 and outward to port 1 of cylinder D. As the piston rod advances, air is exhausted from port 2 of the cylinder through the cylinder port 1 and exhaust port of valve C. As the piston rod in the cylinder approaches the intermediate point X_1, the operator releases the handle of valve C,

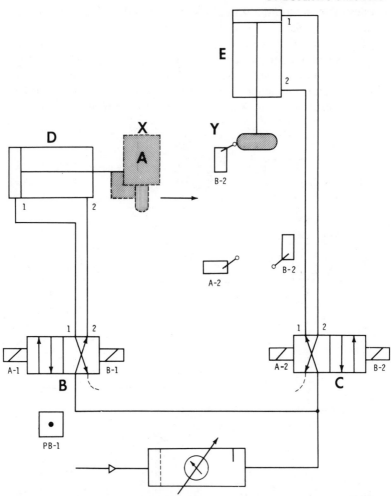

Fig. 7. Schematic diagram of a circuit designed to move a workpiece "around a corner," or from position X, to position Y, to position Z.

allowing the spool in the valve to center, closing all ports in the valve. To retract the piston, the operator shifts the handle to the opposite position, and air is directed from the inlet port to cylinder port 1 and onward to port 2 of cylinder *D*. Then the piston retracts. This device may be used on a stacking mechanism, a transfer mechanism, a press, or any device that is used where it

367

is desirable to stop the piston in a position other than the extreme end of the stroke.

In moving a workpiece "around a corner," or from position *X*, to position *Y*, to position *Z*, a more complicated circuit (Fig. 7) is required, because two complete movements are involved. The operator momentarily depresses electric push button *PB-1* which causes solenoid *A-1* to be energized momentarily and the

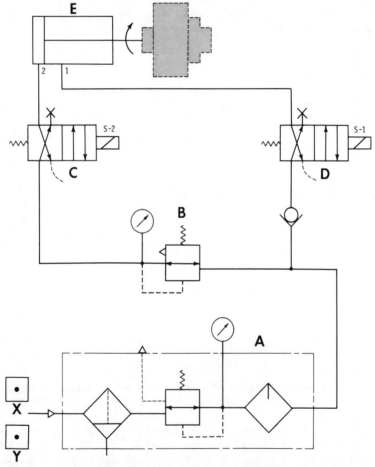

Fig. 8. Schematic diagram of a circuit designed for a two-pressure operation, such as a power chucking operation in which a high clamping pressure is used for the roughing operation, and a low clamping pressure is used for the finishing operation.

spool of valve *B* to shift to direct air from the inlet port to port 1 of the solenoid-operated valve *B* and onward to port 1 of cylinder *D*. The piston of cylinder *D* begins to advance, and the workpiece is moved from position *X* toward position *Y*. As position *Y* is approached, a swing-type trip mechanism on the piston rod of cylinder *D* contacts the limit switch A_2. A swing-type trip mechanism acts only in one direction; therefore, the trip mechanism depresses the actuator of the limit switch as it passes over it in one direction, but it does not depress the limit switch as it moves in the opposite direction. When the limit switch A_2 is actuated, it energizes the solenoid *A-2* of valve *C* and shifts the spool in valve *C*. Air is directed from the inlet to port 1 of valve *C* and then to port 1 of the cylinder *E*. The workpiece is moved from position *Y* toward position *Z*. As the workpiece approaches position *Z*, the trip mechanism on the piston rod of cylinder *E* contacts the limit switch B_2. The solenoid *B-2* of valve *C* is energized, and the spool of valve *C* shifts to its original position. Air is directed from the inlet port to port 2 and then to port 2 of cylinder *E*. The piston rod of cylinder *E* retracts, and the second trip mechanism on the piston rod of cylinder *E* (a swing-type) momentarily contacts the limit switch B_1. This momentarily energize the solenoid *B-1* of valve *B* and the spool of valve *B* shifts to its original position. Air is directed from inlet port to port 2 of valve *B* and then to port 2 of cylinder *D*. The piston of cylinder *D* retracts to its original position, completing the cycle.

Other types of four-way valves can be used in this circuit. Two sequence valves may be used to eliminate one four-way control valve. When sequence valves are used in air circuits, problems may be encountered if the line pressure fluctuates.

An air circuit designed for a two-pressure operation is used in conjunction with power chucking equipment (Fig. 8). Power chucking equipment is used on lathes, grinders, and boring mills, and it is an important facet of production operations. The chief advantage of the two-pressure system is that a high clamping pressure can be used for the roughing operation and a lower clamping pressure can be used for the finishing operation, thereby reducing the possibility of distorting the workpiece during the finishing operation.

In the circuit (see Fig. 8), the operator loads the workpiece into the chuck and depresses the hold button *X*. This energizes the solenoid *S-1* of the four-way valve *D*, and the spool shifts, directing the air to port 1 of the rotating cylinder *E*. The piston of cylinder *E* retracts, under high pressure, and the jaws of the chuck close on the workpiece, gripping it securely. The roughing operation is performed as the spindle revolves. The operator then depresses the hold button *Y*, and the solenoid *S-2* of valve *C* is en-

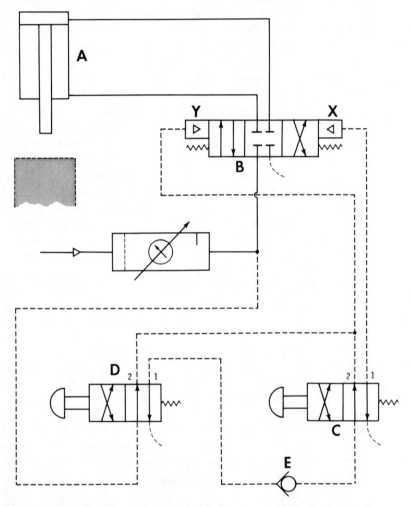

Fig. 9. Schematic diagram of a circuit that can be used when it is desirable to actuate the operating system by means of dual controls.

ergized. The spool of valve C shifts, and low-pressure air passing through the regulating valve B is directed to port 2 of the rotating cylinder E; this places low-pressure air on the blind end of the piston and high-pressure air on the rod end. The clamping force on the workpiece is reduced for the finishing operation, because pressure is exerted on both sides of the piston. It is not necessary to stop the spindle of the machine to accomplish this operation.

After the finishing operation is completed, the operator stops the spindle and releases the hold button X. This releases the solenoid S-1. The spring shifts the spool to its original position and the inlet port of valve D is blocked. Air in the blind end of the cylinder E advances the piston, and the chuck jaws are opened, releasing the workpiece. The hold button Y is then released; this releases the solenoid S-2, the spring shifts the spool to its original position, and the inlet port of valve C is blocked. Air in the blind end of the cylinder is directed to the exhaust port, and the cycle is completed.

Many industrial plants demand that the operating system be actuated by dual controls when there is a possibility of an operator's hands being mangled by a fast-moving piston rod or ram of a cylinder. This type of circuit is shown in Fig. 9.

In the circuit (see Fig. 9) the operator sets the workpiece directly beneath the piston of the cylinder A, and depresses the actuating handles of valves C and D. Air flows from the inlet port of valve D through port 1, through the check valve E, and then to the inlet port of valve C. From valve C, air is directed from port 1 to the pilot connection X of valve B. The spool of valve B shifts, and air is directed to the blind end of cylinder A, and the piston advances at a rapid rate. If the operator removes his hand from valve C during the advance of the piston of cylinder A, the line connecting port 1 of valve C and port X of valve B opens to exhaust, and the spool of valve B centers, thus stopping the movement of the piston of the cylinder A. If the operator removes his hand from the valve D during the advance of the piston of cylinder A, air entering the inlet port of valve D is directed to port 2, onward to port 2 of valve C, and then exhausted. Air that is trapped by the check valve E exhausts quickly through the special piston in valve B, and the spool of valve B centers, thus stopping the advance of the piston of cylinder A. To return

371

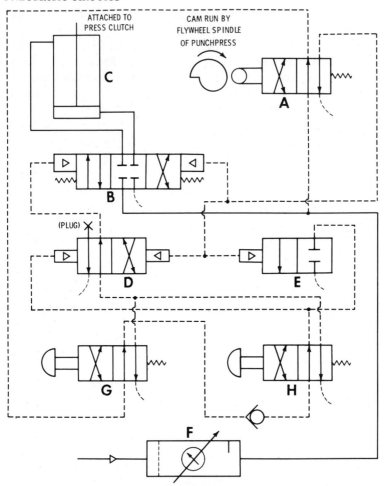

Fig. 10. Schematic diagram of a circuit that can be used on a punch press to actuate the press clutch.

the piston, it is necessary to remove both hands from the actuators of valves *C* and *D*. The circuit cannot complete the cycle if the actuator of either valve (*C* or *D*) is tied down.

Fluid power equipment can be used in conjunction with many types of mechanically actuated mechanisms, such as punch press feeds, bar feeds, transfer mechanisms, etc. A typical circuit for

372

fluid power equipment that is used on a punch press to actuate the press clutch is shown in Fig. 10.

When all the valves (see Fig. 10) are in their normal position and the cam, which is run by the flywheel spindle of the punch press, is not contacting the cam roller of the cam-operated valve *A*, air flows through valves *G* and *H* to the air pilot connection of

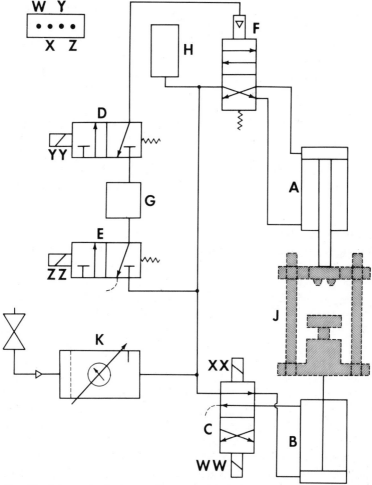

Fig. 11. Schematic diagram of a circuit designed to deliver a hammer-like blow on a workpiece and to provide safety to the operator from a fast-moving piston rod.

373

valve D. This causes the spool of the valve to shift. Then the operator depresses the push buttons of valves G and H; air flows to the chamber of valve B, and the spool shifts. Air from unit F is directed through valve B to the rod end of the clutch cylinder C, and the piston of the cylinder retracts.

When the cam on the flywheel spindle depresses the cam roller of valve A (see Fig. 10), the air is directed to the pilot connection of valve D, the pilot connection of valve B, and the pilot connection of valve E. The spool of valve D shifts, and the air in the chamber of valve D exhausts through valve E. The air in the chamber of valve B exhausts through valve D. Air in valves G and H exhausts through valve A. The air entering the inlet port of valve B is directed to the blind end of cylinder C, and the piston advances.

To begin another cycle it is necessary to release the push buttons of both valve G and valve H, so that the spool of valve D can be repositioned (see Fig. 10). This action requires the operator to keep both hands occupied during the advancing and closing actions of the press. Some applications require the piston of a cylinder to move at a very rapid rate in order to produce a hammer-like blow on a workpiece. The circuit (Fig. 11) not only produces a hammer-like blow from the cylinder piston but also provides safety to the operator from the fast-moving piston rod.

In the circuit shown in Fig. 11, the operator places the workpiece in the air-operated collet chuck J, and momentarily depresses the push button W. This energizes solenoid WW of valve C, and the spool shifts, directing air to the rod end of cylinder B, and the jaws of the collet chuck J close. The workpiece is held securely. Then the operator depresses push buttons Y and Z to energize solenoids YY and ZZ, and air is directed to the pilot connection of valve F. The spool of valve F shifts, and the large volume of air under pressure from the surge tank H and from the source is directed through valve F to the blind end of cylinder A. The piston advances at a very rapid rate, and work is performed on the workpiece. The operator releases push buttons Y and Z which, in turn, cause air to be released from the pilot section of valve F. When the spool of valve F is shifted by spring pressure, air is directed to the rod end of cylinder A, and the piston retracts. The operator momentarily depresses push

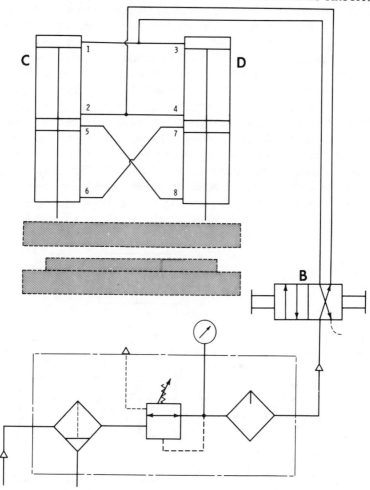

Fig. 12. Schematic diagram of a circuit that employs an air-hydraulic cylinder to synchronize piston movements.

button X which energizes solenoid XX of valve C; the spool of valve C shifts, and air is directed to the blind end of cylinder B. The jaws of the collet chuck open, and the operator removes the completed workpiece.

Synchronization of piston movement by means of air cylinders is impractical, but air-hydraulic cylinders provide an excellent

375

means of providing this type of synchronization. The air-hydraulic cylinders use compressed air for power, and hydraulic oil for control. In the circuit shown in Fig. 12, the operator places the workpiece on the platen of the press, and shifts the handle of the two-position four-way valve *B*. Air is directed to ports 1 and 3 of the air-hydraulic cylinders *C* and *D*. The pistons of these cylinders advance as oil is forced from port 6 to port 7 and from port 8 to port 5. This action keeps the pistons of cylinders *C* and *D* synchronized. When the press is closed and the work is completed, the operator shifts the handle of valve *B* to its original position, and air is directed to ports 2 and 4. The pistons of cylinders *C* and *D* are retracted in unison as oil is directed from port 5 to port 8 and from port 7 to port 6. This system is inex-

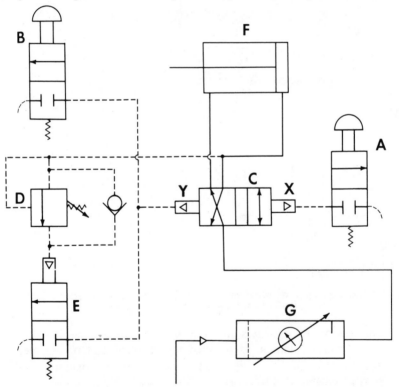

Fig. 13. Schematic diagram of a circuit designed to operate an air cylinder that is required to impart considerable force in some applications.

pensive to set up, is quite accurate, and is adaptable to production methods.

Since air cylinders are capable of imparting considerable force, extensive damage can result to the tool or the workpiece if the force is applied prematurely. The circuit shown in Fig. 13 can be used for many applications.

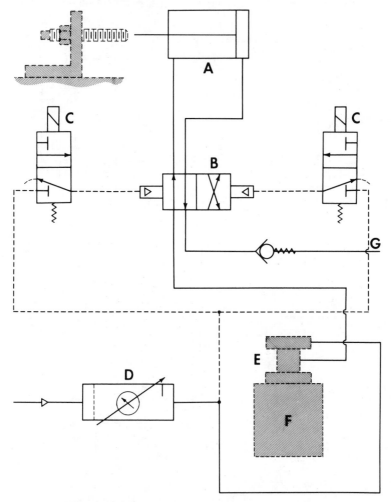

Fig. 14. Schematic diagram of a circuit that uses air valves to control the main hydraulic valve which directs a viscous fluid.

When the operator momentarily trips the actuator of the two-way valve A, air bleeds from the pilot chamber of the four-way valve C (see Fig. 13). The spool of valve C shifts because of the unbalanced condition. Air is directed from the inlet port of valve C to the blind end of cylinder F. A takeoff to the sequence valve D is placed between valve C and cylinder F. If the full line pressure—caused by a jam or other malfunction—or a pre-

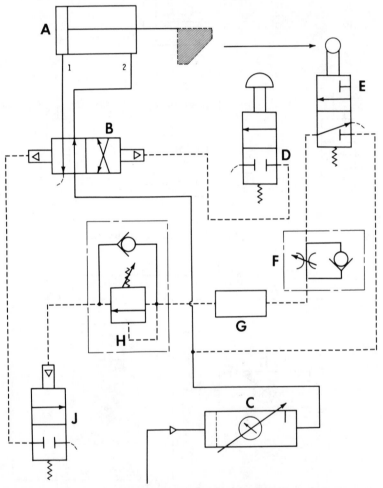

Fig. 15. Schematic diagram for a circuit designed for applications in which a time delay is desirable.

determined set pressure is applied suddenly against the blind end of the piston of cylinder F as the piston is advancing, the sequence valve D opens and actuates the two-way valve E. This allows air to bleed from the chamber of valve C, and the spool shifts. Air is directed immediately to the rod end of the cylinder, and the piston retracts.

Normally, when the piston of cylinder F reaches the end of its forward stroke, pressure in the line between valve C and cylinder F builds up to full line pressure, and the valve D opens. This causes the spool in valve C to shift, eliminating the need for valve B. In applications where there is a possibility of an exceptionally high external force contacting the rod during the advance stroke, it may be desirable to set the opening pressure of valve D above the full line pressure. In this instance, valve B would be required in the circuit.

The metering of specific amounts of certain fluids often becomes a problem, especially if the fluid has a high viscosity and must be measured quickly. In the circuit shown in Fig. 14, air valves are used to control the main hydraulic valve which directs the viscous fluid. This circuit functions well and it is not expensive. A simple external adjustment controls the quantity of fluid that is to be dispensed. The same quantity of fluid is dispensed on either the advance stroke or the return stroke of the piston, because a double-end rod cylinder A is used.

In the circuit, air is directed through the regulator-filter-lubricator unit D to the valves C and to the air-operated grease pump E. Grease, under pressure, is directed to the four-way valve B. When the solenoid of either of the valves C is energized, air pressure shifts the spool of valve B, and grease is directed to cylinder A. As the piston of cylinder A advances, grease which is stored in the opposite end of cylinder A is forced from the cylinder and directed through valve B to the hose G.

This system can be set up for automatic operation. When the part that is to be filled contacts a designated point, one of the two valves marked C functions, thereby making the cycle automatic. This type of system can be used to fill crankcases with oil, to fill transmissions with grease, and for many similar applications.

A time-delay action is sometimes desirable. This type of circuit is shown in Fig. 15. To operate the circuit, the hand-operated air valve *D* is depressed to shift the four-way air control valve *B*, which applies air to cylinder port 1 of the air cylinder *A*. The piston of cylinder *A* moves outward, until the cam on the end of the piston rod operates the cam on the air valve *E*, directing air to the air speed-control valve *F*. An adjustable orifice in the air speed-control valve controls air flow into the air receiver tank *G*. The tank is filled with air from the air speed-control valve, until the pressure setting of the sequence valve *H* equals the spring setting of the valve. When the air sequence valve opens, the air control valve *J* shifts positions to operate the four-way air control valve *B*. With this valve in the shifted position, pressure to port 2 of cylinder *A* returns the piston to a closed position. The length of time delay is controlled by the flow of air into the air receiver and by the size of the air receiver tank.

In designing and installing air circuits, it should be remembered that: (1) air is compressible; (2) air line pressure in an industrial plant generally does not exceed 90 psi; (3) air should be regulated, filtered, and lubricated; and (4) air causes exceptionally quick movements of components, thereby making safety controls mandatory in many installations.

REVIEW QUESTIONS

1. Where should the lubricator be located in a pneumatic system?

2. Draw a simple circuit, using *ANS* Graphic Symbols, showing where the speed control should be located if the piston of the cylinder is to be controlled on the return stroke.

3. Draw a circuit, using *ANS* Graphic Symbols, showing the use of a sequence valve.

4. Draw a circuit, using *ANS* Graphic Symbols, showing the use of a double-acting cylinder.

5. Draw a circuit, using *ANS* Graphic Symbols, showing how to obtain two different pressures in a circuit.

6. Draw a circuit, using *ANS* Graphic Symbols, showing how a rotating cylinder is used in a system.

CHAPTER 19

Hydraulic Circuits

Hydraulic circuitry is not a mystery. If all the movements that are possible in the use of cylinders, motors, and actuators are analyzed, it will be discovered that a study of hydraulics necessarily involves more details than a study of pneumatics. Many of these details have been covered in the preceding chapters, but some of them should be reviewed. In hydraulics, the entire system is usually considered, because central hydraulic systems are in the minority. In pneumatics, only the portion of the system that operates a machine, fixture, conveyor, or similar device is considered; the air compressor is not considered, because it is usually sized to furnish air pressure for the entire plant. Generally, the pneumatic circuit designer is not familiar with the size of the compressor that is being employed. In hydraulics, the designer must select a power source that is large enough to move the cylinders or other motivating means at a speed that fulfills the requirements.

In hydraulics, provisions must be made to handle the exhaust fluid. In pneumatics, the exhaust air can be directed to the atmosphere; however, in hydraulics, provisions must be made to recapture the exhaust fluid by returning it to the reservoir. If the exhaust fluid is not returned properly, pressure that may cause malfunction of the controls may be created. Heat that is created within the hydraulic system becomes an important factor, whereas it is rarely a problem in the pneumatic system.

In hydraulics, the selection of proper fluids may be a problem. Although the air lines often contain foreign matter, proper filtration usually takes care of even the worst condition in the pneu-

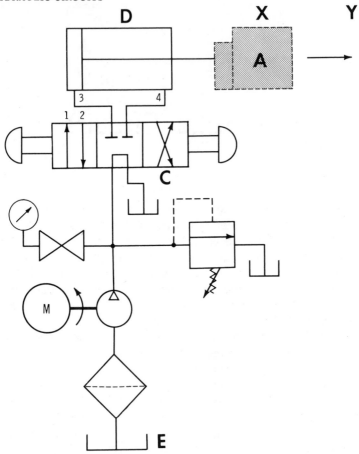

Fig. 1. Schematic diagram of a circuit designed to move a piston at a nearly constant speed.

matic system. A number of basic hydraulic circuits are discussed in this chapter.

A circuit designed to move an object *A* from position *X* to position *Y* at a nearly constant speed is shown in Fig. 1. It may be noted in this circuit that it is necessary to show the power source. The operator loads the object *A* at position *X* and shifts the handle of the three-position open-center four-way valve *C*, and oil, under pressure, is directed from the power unit *E* through the inlet port of valve *C* to port 1 and then to port 3 of cylin-

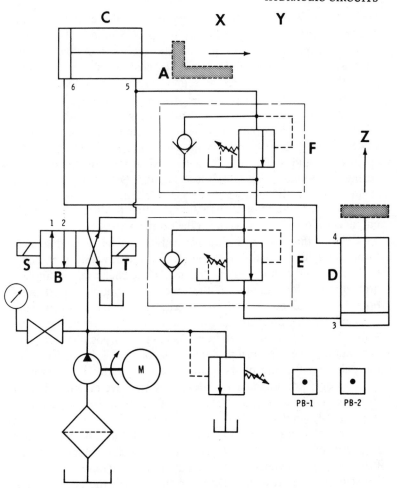

Fig. 2. Schematic diagram of a circuit that uses two sequence valves to perform a sequence operation.

der D. As oil pressure forces the piston forward, oil is exhausted from port 4 to port 2 of valve C and then to the reservoir of the unit E. The piston and piston rod of cylinder D cause the object to move to position Y. Then the operator shifts the handle of valve C to the extreme opposite position, and oil, under pressure, is directed from the inlet of valve C to port 2 and then onward to port 4 of cylinder D. The piston of the cylinder begins to retract, and oil is exhausted through port 3 to port 1 of valve C and through the exhaust port to the reservoir of the unit E. The piston rod of cylinder D retracts to position X. The operator shifts the handle to neutral or center position, and the cycle is completed. By placing the handle in neutral position, the oil is directed to the reservoir at no pressure during the stand-by period; this keeps heat from building up in the system. If a two-position valve, rather than a three-position valve were used, the oil, under pressure, would spill through to the exhaust of the relief valve during the stand-by period, thus creating heat. This does not mean that it is impossible to devise a circuit that uses a two-position valve and unloads the system during the stand-by period, but an additional valve must be added to the system.

A circuit that uses two sequence valves can be devised to perform a sequence operation (Fig. 2). Sequence valves perform well in hydraulic applications, unless an external force is applied prematurely—for example, if the object A is moving from position X to position Y and the object is caught in the moving mechanism, the pressure builds up quickly, the sequence valve F opens, and the movement from position Y to position Z occurs prematurely. Under normal conditions, object A is loaded onto the pallet on the end of the piston rod of cylinder C at position X. The operator momentarily depresses the electric push button $PB-1$, which causes the solenoid S of valve B to be energized momentarily, and the spool of valve B is shifted. Oil is directed from the inlet port to the cylinder port 1 of valve B and onward to port 6 of cylinder C. The piston advances and oil is exhausted from port 5 of cylinder C to port 2 of valve B and out the exhaust port to the reservoir. As the piston moves the object to position Y, pressure builds up in the line from the pump to port 6 of cylinder C, the passage through sequence valve E opens, and the oil is directed to port 3 of cylinder D. The piston and piston rod of cylinder D advance and move the object A from position Y to position Z as the oil is exhausted from port 4 of cylinder D, passes through the check valve in the sequence valve F to port 2 of valve B, and out the exhaust port to the reservoir. The operator depresses push button $PB-2$ causing solenoid T to energize momentarily, and shift the spool of valve B to its original position. Oil is directed to port 2 of valve B and then to port 5 of cylinder C, and the piston retracts. Pressure builds up in the line from the pump to port 5 of cylinder C, the passage through the sequence valve F is opened, and oil passes to port 4 of cylinder D. The piston retracts to complete the cycle.

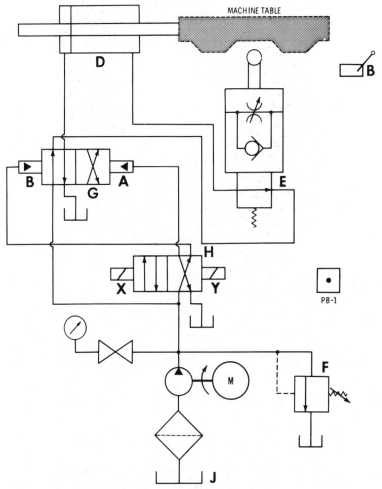

Fig. 3. A schematic diagram of a circuit that is used to decelerate an object while work is performed on it at certain points, as the object is moved from one position to another.

If it is not necessary to retract the pistons of cylinders C and D in sequence, sequence valve F may be omitted. If it is necessary to retract the piston of cylinder D before the piston of cylinder C is retracted, the piping should be run directly from port 2 of valve B to port 4 of cylinder D. A tee should be placed in this line with the sequence valve F connected to it. Then a line should be run from sequence valve F to port 5 of cylinder C.

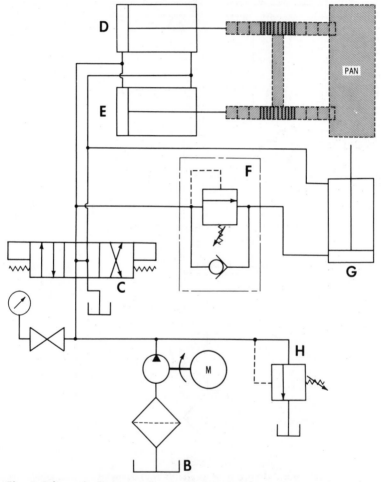

Fig. 4. Schematic diagram of a circuit designed to synchronize the movements of the pistons of two separate cylinders.

A circuit in which the object A is decelerated at certain points while work is performed, as the object moves from one position to another, is shown in Fig. 3. To operate the circuit, the operator places the workpiece on the machine table which is equipped with a magnetic chuck. The operator momentarily depresses push button PB-1, which energizes the solenoid X of valve H. This

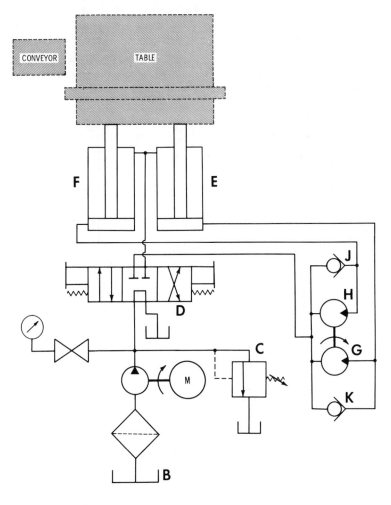

Fig. 5. Schematic diagram of a circuit designed to illustrate the synchronization of the pistons of two separate cylinders.

directs pilot pressure to the pilot A of valve G. The spool shifts, allowing oil to flow to the blind end of the feed cylinder D. The feed cylinder piston rod moves the machine table forward at a rapid rate, until the cam on the feed table contacts the cam roller of the cam-operated speed control valve E. As the cam roller is depressed, the exhaust oil flow is shut off, and it meters through the speed-control portion at a speed determined by the needle setting. When the roller rides off the cam, the cylinder operates at full speed. When the piston rod of cylinder D reaches the end of its stroke, the limit switch B is contacted, energizing the solenoid Y, shifting the spool of valve H to its original position, and directing oil to the pilot B. The spool of valve G shifts to direct

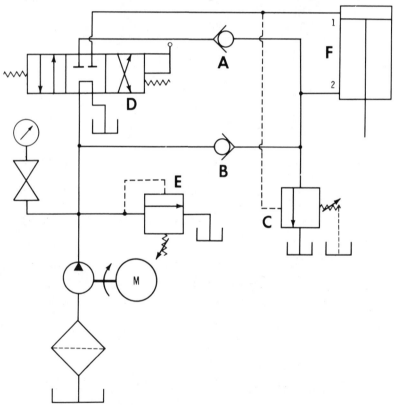

Fig. 6. Schematic diagram of a circuit designed to regenerate a portion of the oil to help speed up the operation of the cylinder.

oil to the rod end of cylinder D, whereupon the piston and rod retract the table at a rapid rate. The operator unloads the workpiece, reloads, and is ready for the next operation.

The pilot control valve H controls the action of valve G. On equipment that provides a central station for the operator, push buttons may be used to save on installation costs. Skip-feed valves are often used on grinders and other large machines.

A rack-and-pinion arrangement may be used to synchronize the movements of the pistons of two separate cylinders (Fig. 4). In many applications, the pistons of two cylinders must move in unison to perform a given task. After the tray has been filled with workpieces, the operator shifts the handle of the master three-position, open-center, four-way valve C, allowing oil to flow to the blind ends of feed cylinders D and E. The force of the oil moves the pistons forward. The ends of the piston rods are equipped with racks, and they ride on a common pinion which enables the cylinder pistons to move forward together. The tray moves forward on the conveyor table.

At the end of the stroke, pressure builds up in cylinders D and E, the sequence valve F opens, and oil flows to the blind end of the transfer cylinder G. The piston and rod of the transfer cylinder move forward, removing the tray from the conveyor table.

The operator shifts the handle on valve C, and oil flows to the rod end of all three cylinders, returning their pistons to their original positions. The operator must keep a hand on the handle of the control valve C at all times during the movement of the cylinders, or the springs will center the valve, causing the cylinders to stop and the pressure to exhaust to the tank.

Synchronization of the movement of two cylinders is illustrated in Fig. 5. In the circuit, the workpiece moves onto the table from the conveyor. The operator shifts the handle of the three-position, spring-center, four-way control valve D. Oil flows to the fluid motors G and H, which separate the oil equally. Oil then flows to the blind ends of the pusher cylinders E and F.

Oil pressure advances the pistons of the pusher cylinders at the same rate until the end of the stroke is reached. Then the operator shifts the master control valve D, oil flows to the rod end of the pusher cylinders E and F, and the pistons return to their original positions. The operator releases the valve handle, allowing the piston to center, and the oil flows to the sump at zero pressure during the stand-by period.

Regeneration of a portion of the oil in a system helps to speed up the operation of the cylinder (Fig. 6). When the spool of the four-way valve D is in neutral position, oil from the pump is returned to the reservoir at low pressure. The operator shifts the handle of valve D, and oil is directed to port 1 of cylinder F for the advance stroke. The oil from port 2 of cylinder F cannot exhaust or return through valve D due to the check valve A. Oil is then forced through check valve B and into the inlet port of valve D. The pump volume and the volume of oil forced from port 2 of cylinder F is then forced into port 1 of cylinder F at a pressure much lower than the relief valve setting for the pump. When the piston rod of cylinder F contacts the workpiece, pressure builds up in the pilot line of valve C. As the preset opening pressure is reached, the passage in valve C automatically opens, and the oil that is flowing from port 2 of cylinder F is returned to the reservoir under no pressure. This allows full line pressure to be applied to the upper side of the piston of cylinder F during the work cycle. At the completion of the work cycle, the operator reverses the spool of valve D to the opposite position. Oil is directed through check valve A to port 2 of cylinder F, and the piston is retracted as the oil is exhausted from port 1 to the tank port in valve D. This action releases pilot pressure to valve C, and the passage in valve C is closed. In this circuit, the flow of oil is regenerated through valve D; therefore, the capacity of valve D must be large enough to handle the capacity of the pump, plus the quantity of oil regenerated from the rod end of cylinder F.

Some applications require several different forces to complete an operation. These forces may be obtained by the use of cylinders of different sizes, but this is not always possible, because the forces required for a given job may be so small that a single cylinder of suitable bore size is unavailable. A circuit designed to change operating pressures in various parts of the system is diagrammed in Fig. 7. This system is used to press fit (light press) cylinder liners into cylinder bodies.

The main operating pressure of the system is controlled by the relief valve N on the power unit. The pressure reducing valve

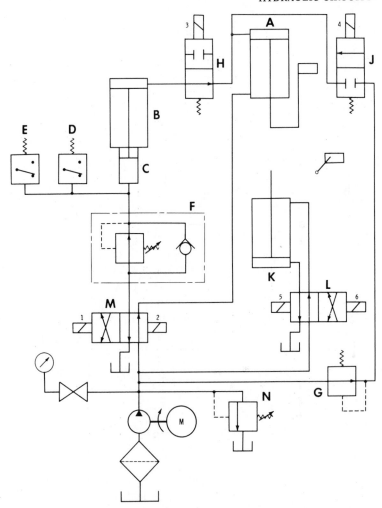

Fig. 7. Schematic diagram of a circuit designed to change operating pressures in various parts of the system.

F controls the pressure to cylinder C. The pressure reducing valve G controls the pressure to the upper end of cylinder A. To operate the circuit, the operator loads the liner into a fixture and momentarily depresses push button X. This momentarily energizes solenoid 1 of valve M, and the spool of valve M shifts.

391

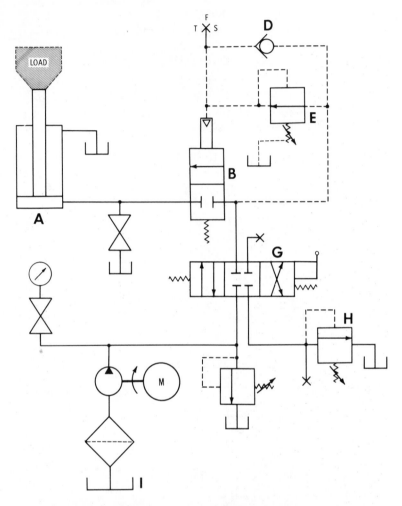

Fig. 8. Schematic diagram of a circuit designed to prevent dropping a load that is being lifted or tilted during a power failure.

directing oil to the valve F and onward to the blind end of cylinder C. The pressure switches D and E are placed in this line. A green indicator light is connected to pressure switch D, and a red indicator light is connected to pressure switch E. A satisfactory press fit is indicated by the green light (switch D), and a press fit that is too tight is indicated by the red light (switch E). The passage in the two-way valve H is normally open. As the piston of cylinder C begins to move low-pressure oil in cylinder B and since there is a large difference in piston area between cylinders C and B, the oil is directed to the top of the pressing cylinder A, and the piston of cylinder A advances. When the limit switch S is contacted and if the press fit is too tight (red light has been "ON"), solenoids 2 and 5 are energized, the piston of cylinder A is retracted, and the piston of cylinder K advances at the pressure level set by valve N. If the green indicator light is "ON" when switch S is contacted, solenoid 3 closes valve H, solenoid 4 opens valve J, and the pressure set by valve G is applied to the top side of cylinder A to finish the operation. When the operator momentarily depresses push button Y, solenoid 2 is energized, and the spool of valve M is returned to normal position. Oil is directed to the rod end of cylinder A, and the piston retracts, completing the cycle.

Various hydraulic safety systems are used; the system that is used depends largely on the operation that the designer is endeavoring to protect. The circuit shown in Fig. 8 may be used where a load is to be lifted or tilted. If there is a power failure, the load will not drop—as in the tilting of a furnace to pour molten metal, for example, where a power failure may cause serious damage. The operator shifts the handle of valve G to move the piston of cylinder A to the "UP," "DOWN," or "JOG" position. If either the electric motor or the pump on the power unit I fails during the motion of the piston of cylinder A, a pressure loss occurs. The pilot-operated valve B is connected to the blind-end port of cylinder A. Pilot pressure is taken from the line at point S. Pressure in this line is required to keep the passage in valve B open. If a pressure failure occurs, the passage in valve B will close to keep the piston of cylinder A from retracting and causing damage.

To return the piston of cylinder A to its retracted position, the operator manually bleeds the cylinder through the shutoff valve C. Valve E is used when the pilot chamber of valve B is not suitable to accept the working pressure. Valve D is used to exhaust pilot pressure when a pressure failure occurs. In instances where the pilot chamber of valve B can accept the working pres-

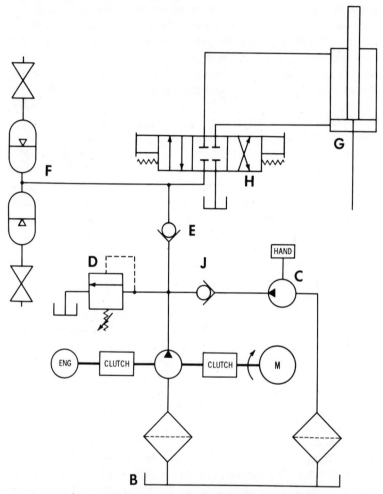

Fig. 9. Schematic diagram of a circuit designed to provide a reserve power supply in event of an electric power failure.

sure, valves D and E can be deleted. Also, in instances where "jogging" is not required, valve G can be replaced by a two-position three-way valve.

The test station F is used only for the setting of valve E and is deleted if the pilot chamber of valve B can accept the working pressure. Valve H is set at a level that maintains only enough back pressure to keep valve B open, in order to return the piston of cylinder A under normal conditions.

By revamping the circuit, cylinder A can be converted to a double-acting cylinder with an identical automatic locking safety feature. It should be noted that valve B does not depend on electrical operation, thereby eliminating possibilities of solenoid failure.

It may be necessary to provide for a reserve power supply in event of a failure of the primary power source, such as an electrical failure (Fig. 9). In some applications, a movement must take place, even though a complete power failure occurs. Some of these applications are: to close the large gates on a power dam, to actuate large gate-type valves, and to remove workpieces from gas-fired furnaces. In normal operation, the clutch connected to the electric motor is engaged, and the fixed-delivery pump is actuated by the electric motor to deliver oil to the system. If an electric power failure occurs, the clutch on the side of the electric motor is disengaged, and the clutch on the side of the internal combustion engine is engaged after the engine is started. Then the pump is actuated, and oil is delivered to the system. The accumulators F are placed in the system as a reserve source of oil under pressure.

If both the electric motor and the internal combustion engine fail, the hand pump C can be used to supply pressure to the system. Although the hand pump is a slow method of furnishing power, it is very reliable. The check valves J and E are used, so that oil, under pressure, cannot escape when the engine or motor is stopped.

Hydraulic circuits are exceptionally well adapted for controlling tool movements on lathes, grinders, and other types of precision machines. In the diagram (Fig. 10), a lathe feed arrangement is controlled by cylinders E and G. Cylinder G moves the toolholder into the cutting area, and cylinder E moves the toolholder into the cross-feed cutting operation. Both cylinders approach the cutting area at a rapid rate, until the cam roller on the cam-operated flow control valve D is depressed. The movement of the cylinders then enters the feed control operation. When the cam follower contacts the template, the cylinder E holds its position, and the piston of cylinder G advances. As the piston of cylinder

395

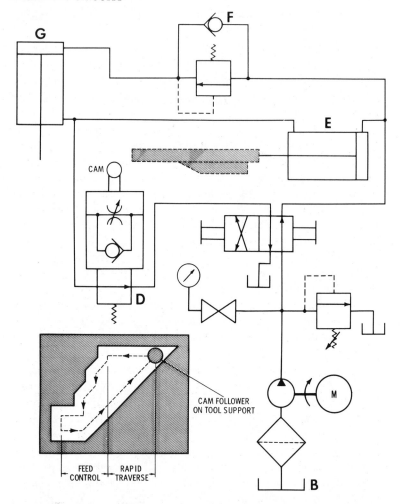

Fig. 10. Schematic diagram of a circuit designed to control tool movements on lathes, grinders, and other types of precision machines.

E advances, cylinder G holds position. This is possible because reduced pressure to cylinder E is provided through the pressure reducing valve F.

Hydraulic components should be secured well, because most hydraulic circuits operate in the higher pressure ranges. Circuits should be designed to keep the oil at a suitable operating temperature, to eliminate restrictions in piping, to provide ample oil reservoirs, and to protect the system from dirt and heat.

REVIEW QUESTIONS

1. Show, using *ANS* Graphic Symbols, how a heat exchanger should be connected to a power unit.

2. Draw a circuit, using *ANS* Graphic Symbols, of a power unit with two pumps.

3. Draw a circuit, using *ANS* Graphic Symbols, in which a hydraulic pilot-operated valve is used to actuate a cylinder.

4. Draw a circuit, using *ANS* Graphic Symbols, in which a reducing valve is used.

5. Draw a circuit, using *ANS* Graphic Symbols, in which an actuator is used.

6. Draw a circuit, using *ANS* Graphic Symbols, in which a hydraulic motor is used in conjunction with a hydraulic cylinder.

CHAPTER 20

Fluid Circuit Failures

In the preceding chapters, fluid power components have been discussed in detail. New component designs are coming off the drafting boards each day. With the tremendous amount of applications and demands for fluid power devices, the manufacturer of fluid power components has become a part of one of the fastest growing industries. The next decade will surely lead to a number of improvements in fluid power components, but the basic principles will undoubtedly remain the same.

COMMON CAUSES OF FAILURE

The causes of component failures have been studied. These facts should always be kept in mind when servicing fluid power devices. Some of these points should be reviewed.

Dirt

Without doubt, dirt causes more components to fail than any other single cause. Dirt also includes foreign substances.

In a pneumatic system, dirt and foreign substances score the honed cylinder tubes, precision-finished valve liners and valve seats, ground and polished piston rods, valve stems, and other precision parts. In pneumatics, the foreign matter may be in the form of: pipe scale; lime deposits; thread compound; shavings from pipe threads; corrosive fumes entering the intake of the compressor and being distributed throughout the system; welding spatter caused by carelessness during construction; rust caused either by improper filters or by excessive condensation; sand and dirt in the components caused by removal of the pipe plugs before the components are installed; and deposits on piston rods caused by particles in the air, which can be drawn into the system. Dirt embeds itself in the cylinder cup packings and valve packings, often cutting the packings.

In a hydraulic system, dirt and foreign matter may cause excessive damage to the components, since the fits between the parts are held to very close limits. Dirt not only scores the parts but often causes valve spools to stick and become inoperative. Dirt sometimes becomes lodged between the piston, piston ring, and tube of a hydraulic cylinder, causing the piston ring to be broken. This, in turn, may cause the tube to be badly scored as shown in Fig. 1. Note that the metal appears to be actually scooped out.

Dirt clogs small orifices in valves, causing them to malfunction. Dirt tears the rod packing, and causes excessive external leakage of the fluid. Foreign matter also causes pitting of the piston rods and valve stems. Foreign matter, such as hydrocarbons, may clog intake strainers and cause carbons or cavitation within the pump. Intake strainers have been known to collapse due to a col-

Fig. 1. A scored hydraulic cylinder tube.

lection of foreign matter. Dirt can cause a pump to seize, and the driving means may twist off the pump shaft.

Cutting oils and coolants sometimes get into the hydraulic oil, causing considerable corrosion within the system and failure of the components. Every precaution should be taken to keep these solutions out of the hydraulic system.

Heat

Heat causes considerable trouble to the components of the fluid power system, especially the hydraulic components. Heat may cause valve spools to stick, packings to deteriorate, oil to break down, deposits to cling to the finished surfaces, excessive external and internal leakage, and inaccurate feeds in hydraulic systems. Fluid power systems should be protected from hot blasts. If heat in a hydraulic system is caused by internal conditions, install aftercoolers, and if possible, correct the condition, that is causing the heat. Some of the causes of heat are high ambient temperature, restrictions in hydraulic lines and components, high pressures, and high pressures being spilled through the relief valve.

Misapplication

Misapplication causes many failures of fluid power components. The selection of the incorrect component as to capacity, ability to withstand shock loads, or ability to withstand certain other operating conditions may cause failures. The use of a pneumatic valve for high-pressure oil service is likely to cause trouble. The use of a cylinder with thin cast iron covers for heavy-duty mill applications is almost certain to cause trouble. Misapplication is often a product of misinformation or lack of information on the part of the buyer. It has been found that the buyer often is not willing to divulge to the vendor how the equipment is to be used, because he is fearful that the vendor may learn a trade secret.

Improper Fluids

Use of improper fluids in a hydraulic system may cause failures. Care should be used in selecting the fluid to be used in the hydraulic system. Check with the pump manufacturer for his recommendations. If the oil is satisfactory for the pump, it is likely that it is satisfactory for the other components of the system.

401

As discussed previously, certain hydraulic fluids have detrimental effects on seals, packings, paint, and strainers; if these fluids are to be used, provisions must be made accordingly. Mixing of hydraulic fluids is not recommended as one of the fluids may have a property that is detrimental to the other. Fluids that cause deposits or corrosive action should not be used in hydraulic systems.

Faulty Installation

Faulty installation may contribute to many fluid power system failures. Many instances of faulty installation have been found in various installations. Some of these are:

1. Flow controls are often reversed in the system.
2. Wrong connections are made to directional controls. This can happen on the more complicated circuits where electric valves are involved. The piping and electrical diagrams should be followed closely.
3. Installation of a hydraulic power device so that back pressure is created in the return line to the reservoir. In other words, if it is necessary to push the exhaust oil "uphill" in order to return it to the reservoir, back pressure is created. This causes some of the directional and pressure control valves to malfunction.
4. Installation of hydraulic power devices in a closed pit. This usually causes a heat problem. The oil and pump heat up, and a heat condition is caused in the motor. The power device should be well ventilated.
5. Failure to make drain connections to hydraulic valves is a common cause of failure. When a manufacturer of a control device hangs a tag on a valve port marked "connect to drain," he means exactly that. Many service calls may be eliminated if this suggestion is followed strictly.
6. Installation of piping of inadequate size in either the pressure or the exhaust lines in a hydraulic system may cause trouble. It slows down the action of the system, creates heat, causes malfunction of valves, and creates back pressure. In a pneumatic system, it may cause sluggishness in the action of the components.

Fig. 2. Leaks are a source of trouble.

Fig. 3. Use good housekeeping methods.

7. Cylinders not securely mounted may cause difficulties. Remember that cylinders impart considerable force and if they are not properly mounted, difficulty can be expected. Base-mounted cylinders, especially hydraulic cylinders, should be keyed to relieve the thrust on the mounting screws.

8. If control valves with mounting feet are not mounted on a flat surface, they often cause trouble, as distortion occurs when the mounting feet are securely bolted down.

9. Loose pipe lines are often a source of trouble, especially

403

Fig. 4. What component is missing in the circuit?

Fig. 5. A regulator, a filter, and a lubricator should always be used.

in high-pressure hydraulic systems. Piping should be securely anchored. Strap anchors are often used for this purpose.

10. Lack of protection to the piston rods of cylinders that are installed in dirty atmospheres is another source of trouble. The rod may be scored and dirt may cut the rod packing.

11. Misalignment of piston rods of nonrotating hydraulic cylinders and air cylinders is a source of trouble. Misalignment causes bent piston rods, loss of power, broken covers, worn bearings, scored cylinder walls, and packing leaks.

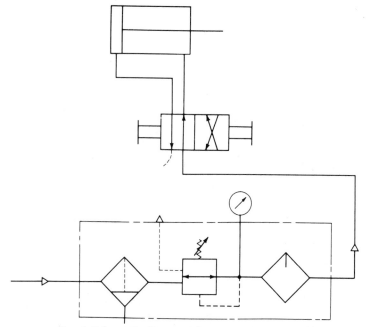

Fig. 6. Schematic diagram of a pneumatic circuit that is protected properly.

12. Leaks around the pipe ports and connections due to improper installation are also troublesome (Fig. 2). In pneumatic systems, leaks are expensive as they often go unheeded for long periods of time. In hydraulics, oil leaks not only are messy, but they also present a real fire hazzard. Always make certain that all pipe connections are tight.

In previous chapters, other causes for failure due to improper or careless installations have been discussed. Careful planning and the ability to produce good workmanship is a prerequisite to a satisfactory installation. Remember that precision equipment, often amounting to several thousand dollars for a single component, is being installed.

Maintenance

Poor maintenance is often a cause for fluid power system failure. A regular maintenance program can reduce failures.

405

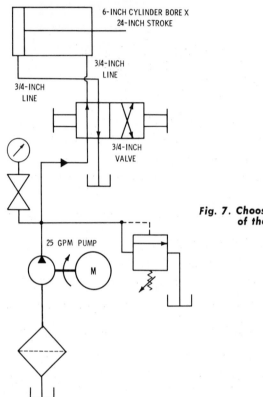

6-INCH CYLINDER BORE X
24-INCH STROKE

3/4-INCH
LINE

3/4-INCH
LINE

3/4-INCH
VALVE

25 GPM PUMP

M

Fig. 7. Choose valves and piping of the correct size.

In a pneumatic system, the lubricators should be filled regularly with a lubricant that is suitable for the pneumatic system. A light spindle oil or a nondetergent oil SAE 10 is generally satisfactory. The filters should be cleaned at regular intervals. Replace worn packings, seals, and other parts to keep the system fully efficient.

In a hydraulic system, the oil should be changed at regular intervals. At that time the system should be cleaned. Check the system for leaks. Change packings and seals when necessary. Clean up oil that is spilled on the floor or on the components. Use good housekeeping methods (Fig. 3).

In either pneumatic or hydraulic systems do not allow dirt to accumulate around the system. Keep it clean.

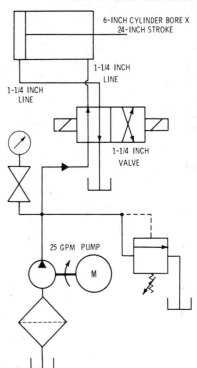

6-INCH CYLINDER BORE X
24-INCH STROKE

1-1/4 INCH
LINE

1-1/4 INCH
LINE

1-1/4 INCH
VALVE

25 GPM PUMP

M

Fig. 8. Schematic diagram of a hydraulic system having valves and piping of the correct size.

IMPROPERLY DESIGNED CIRCUITS

Some circuits are designed incorrectly. This is an important cause of malfunctions in fluid circuits.

A simple pneumatic circuit is shown in Fig. 4. Why will it fail? Note closely that there is no filter, lubricator, or regulator in this system. The absence of these components shortens the life of the other components in the system. The necessity of these three components in a successful pneumatic system has been discussed earlier. Fig. 5 is a comic illustration of this requirement. Also remember to keep the lubricator full of oil and to clean the filter at regular intervals. If the system is used where noise is a factor, install a muffler. See Fig. 6 for the solution of Fig. 4.

A hydraulic circuit is shown in Fig. 7. Why should this system give unsatisfactory service? Note the size of the piping and also

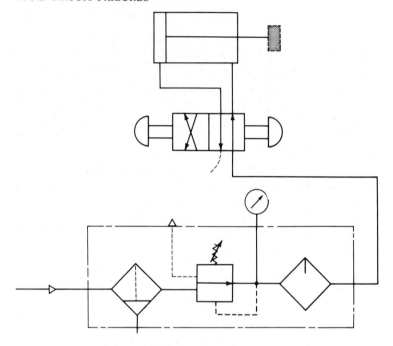

Fig. 9. Schematic diagram of a system lacking a means of speed control.

note the type of the cylinder. Usually, a cylinder with a 2-to-1 rod is able to retract the piston at twice the rate that it moves outward. With a restriction in the valve and in the piping, this is impossible. It should be remembered that in order to remove the oil from the rear of the cylinder, the ports and pipes must be capable of handling oil at twice the rate that it enters at the rear. This is a very common error, and is often overlooked by capable circuit designers. Fig. 8 shows the correct sizes to use.

A pneumatic system that is used for a transfer device is illustrated in Fig. 9. This system fails to give satisfactory service because there is no means of controlling the speed of the outstroke of the piston of the cylinder and no means of relieving the shock at the end of the stroke. Since air is highly expansible, it can move pistons at high velocities, which may create an excessive shock at the end of the piston stroke, especially if the piston

CYLINDER CUSHIONED BOTH ENDS

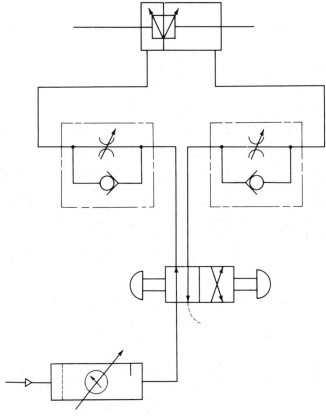

Fig. 10. Schematic diagram illustrating the use of flow controls and cushions to reduce shock.

rod is moving a heavy freely moving load. By the installation of a flow control and cushions on the cylinder, the shock can be reduced greatly (Fig. 10).

A hydraulic circuit is diagrammed in Fig. 11. What could cause this circuit to fail? Note that the power unit is mounted overhead and that the exhaust oil must be forced "uphill" to the reservoir. This creates back pressure on all the valves, which often causes them to malfunction. Keep the unit as low as possible. If necessary, an auxiliary reservoir should be added to col-

409

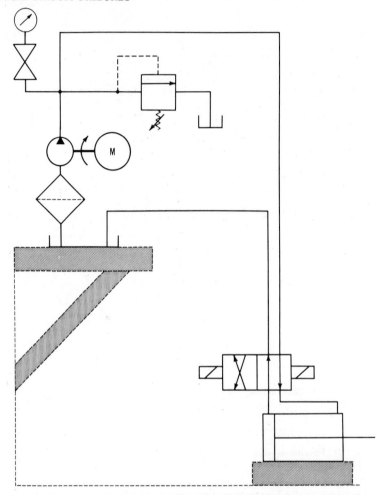

Fig. 11. The power unit should not be mounted at a level higher than the remainder of the system.

lect the exhaust oil; then the oil can be pumped back into the main reservoir.

As shown in Fig. 12, failure to keep air out of a closed hydraulic system renders unsatisfactory results. A jumpy feed breaks tools, causes spoilage of workpieces, and produces vibration. In a closed system, keep the air out; and keep the oil from leaking.

410

Fig. 12. Air should be kept out of a closed hydraulic system.

REVIEW QUESTIONS

1. Explain how dirt may cause trouble in a hydraulic system.

2. Explain how dirt may cause trouble in a pneumatic system.

3. Explain how heat may cause trouble in a fluid power system.

4. How does the use of improper oil cause failure in a system?

5. Explain any cause of faulty installation that you may have seen. What lesson did this case illustrate?

6. Why does air cause trouble in a hydraulic system?

7. What precautions should be used when mufflers are installed in a pneumatic system?

8. What precautions should be taken if a hydraulic system is to be used in an explosive atmosphere?

9. If there is a choice of either hydraulics or pneumatics for the following applications, which would you choose? Explain your answers.

   (a) Fine feed on a machine tool.
   (b) Device for changing the rolls on a steel mill.
   (c) Automatic transfer mechanism for small parts.

411

   (d) Device for providing a hammer-like blow.

   (e) Heavy-duty molding press.

   (f) Device for actuating a power-operated chuck on a heavy-duty lathe.

CHAPTER 21

Record Systems for
Fluid Power Machines

A record system regarding the performance and maintenance of fluid power equipment is essential in modern industrial plants. The amount of fluid power equipment in service determines how elaborate the record system should be. Some of the reasons for keeping these records are: (1) to check the performance of the various components involved; (2) to be certain that critical repair parts are either on hand or readily available; and (3) to establish a preventive maintenance program.

The responsibility for setting up a record system for fluid power components depends on the plant operation. This is the responsibility of the plant engineer in some industrial plants; however, it may be the responsibility of the maintenance superintendent. In still other industrial plants, the tool supervisor may be responsible for the records system. Full-time personnel are required to maintain the fluid power equipment records in some large industrial plants. The person assigned responsibility for the record system should have some concept of the operation of fluid power equipment and the performance of the components.

RECORD OF PERFORMANCE OF COMPONENTS

Many of the large users of fluid power components either purchase or place on test comparable components from a number of fluid power equipment manufacturers. Tests are usually accelerated; the components may be cycled many times more than normal use—over a short period of time. If a group of cylinders are being tested, for example, they are examined periodically to check internal leakage past the piston, leakage around the packing gland assembly, amount of wear in the cylinder walls, rod packing, piston packing and rod bushing, cushioning action, etc. A test of this nature may prove some points, but it does not necessarily prove the ability of the cylinder to perform over a long period of operation. In some applications, a cylinder is required to operate only a limited number of times per month, or even a few times per year, but it is vitally important that it operates satisfactorily each time that it is required to perform. For example, much damage may result if the cylinder that operates a gate on a dam fails to perform when it is needed. In this instance, the packing and seals should not be affected by weather, and the packings cannot have an affinity for the piston rods or cylinder tubes; otherwise, a sticking condition may occur to prevent operation of the cylinders. Malfunction may occur quickly if the packing sticks to the valve stems. Actual operation in the field is required to prove the merit of a component.

A card record system may be used to set up a record of performance (Fig. 1). The record card is set up by machine or fixture numbers. Many cards may be required for a complicated machine, each specifying a component on the machine. It is important to designate the serial number or lot number of the component; manufacturers often change the serial or lot number, but they often retain the original model number.

In the event that repairs are needed, it is important to have the name of the manufacturer of the component and his representative in the area readily available. The installation dates and the repair dates of the components should also be provided on the card. All this information is important in determining how well the component has performed over a period of time.

A cross reference on the card may be valuable in some instances to indicate that the component is used on other machines or fixtures. In an emergency, the component may be removed from a machine that is not being used regularly and substituted for a component that needs repair. Substitution is usually a poor practice, but it may be valuable in an emergency. This is sometimes necessary in an instance where delivery of a special component may require a considerable period of time.

NAME OF COMPONENT: *Four-Way Air Control Valve*
Solenoid Operated 110V-60c

MODEL NO. *16820*   SERIAL NO. *939370 Lot No. CC*

MANUFACTURER: *Ajax Inc.*
Rossburg, Indiana

AREA DISTRIBUTOR: *Industrial Marketing*
Cosco, Indiana   Phone 753-5473

REPAIR KIT NO. *Z-16753*

MACHINE NO.	DATE INSTALLED	DATES REPAIRED	
700	*Jan. 12, 1969*	*1/10/71*	*7/15/73*
863	*June 15, 1969*	*2/15/72*	
990	*Feb. 8, 1971*	*6/10/73*	*6/19/74*
1015	*Oct. 5, 1971*	*7/15/73*	
1132	*May 8, 1974*	*11/20/74*	
1265	*Jan. 8, 1975*		

**Fig. 1. Sample record card that may be used to set up
a record of performance in a card record system.**

The parts list of a component is another important cross-reference item (Fig. 2). The parts list may be attached to the record of performance card (see Fig. 1) if a standard (8 1/2 × 11 inches) sheet of paper is used, but a separate file for the parts lists is recommended if small cards are used.

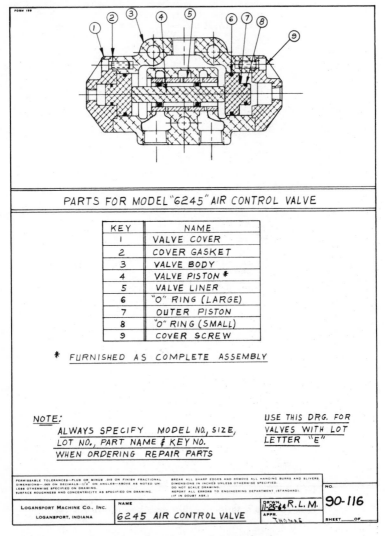

Fig. 2. Parts sheet for an air control valve.

REPAIR PARTS AVAILABILITY

Another card system can be set up and used by the maintenance department to show the number of spare parts that are

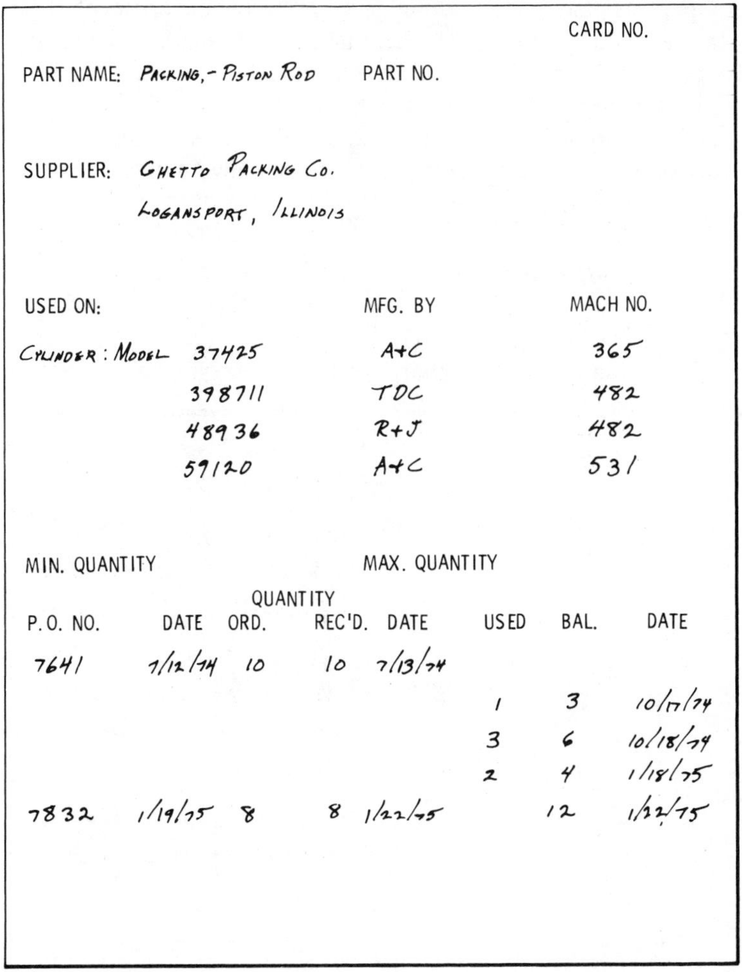

CARD NO.

PART NAME: *Packing,- Piston Rod*   PART NO.

SUPPLIER: *Ghetto Packing Co.*
Logansport, Illinois

USED ON:	MFG. BY	MACH NO.
Cylinder : Model 37425	A+C	365
398711	TDC	482
48936	R+J	482
59120	A+C	531

MIN. QUANTITY   MAX. QUANTITY

QUANTITY

P.O. NO.	DATE	ORD.	REC'D.	DATE	USED	BAL.	DATE
7641	7/12/74	10	10	7/13/74			
					1	3	10/17/74
					3	6	10/18/74
					2	4	1/18/75
7832	1/19/75	8	8	1/22/75		12	1/22/75

Fig. 3. Sample record card indicating the number of spare parts in stock.

in stock (Fig. 3). A running inventory is important in providing spare parts at all times.

In setting up this type of card system, the user may establish his own numbering system; or he may use his suppliers' parts numbers. Fewer cards may be involved if the user sets up his own numbering system, because several manufacturers of like components may have certain items in common. Some of these items are rod packing, seals, bearings, etc. Then these items can be purchased directly from their manufacturer, rather than from the manufacturer of the component.

It is important to set up maximum and minimum stock quantities. Some parts in fluid power components have a shorter life on

PREVENTIVE MAINTENANCE PROGRAM

SYSTEM: HYDRAULIC

MACHINE NO. *1632*

DATE SET FOR REPAIR	DATE STARTED	DATE COMPLETED	NATURE OF REPAIRS
8/15/72	8/16/72	8/19/72	COMPLETE CLEANING OF SYSTEM REPLACED PUMP CHANGED OIL
2/15/73	2/15/73	2/16/73	CHECKED SYSTEM REPLACED CYL. PACK ON MODEL 77642 REPLACED NEEDLE PACKING VALVE 31774
8/15/73	8/15/73	8/18/73	COMPLETE CLEANING OF SYSTEM REPLACED FILTER 774883 INSTALLED NEW STEM ON VALVE 43202H CHANGED OIL
2/15/74	2/15/74	2/16/74	COMPLETE CLEANING OF SYSTEM REPLACED SEAL IN PUMP
8/15/74			

Fig. 4. Sample record card that may be used for a preventive maintenance program.

the shelf than other parts, so it is not advantageous to order a large quantity of these items just to receive a better quantity price. Some types of seals and packings deteriorate with age. Usually, the "name brands" of spare parts should be used as substitutions, because poor performance often makes the unknown substitute parts more expensive.

The card system (see Fig. 3) should be kept current. Valuable time may be lost if the card shows that there is a spare part in stock, and the maintenance man finds that it is missing. This type of card system can be used in conjunction with the parts list shown in Fig. 2; the card number may be placed on the parts list for easy reference. Packings are the chief replacement items; they should be stored away from heated areas, such as near heating pipes and the direct rays of the sun. It is a good practice to store the packings in their original containers. The packings that are small in size are sometimes removed from their original containers, and sealed in glass containers.

Solenoids should also be carried in stock. Extra solenoid coils should be kept available in the voltages that are normally used.

Spare hydraulic pumps, as well as the parts for these pumps, should be carried in stock. Many of these pumps are not serviced easily in the field, and they are returned to the manufacturer for servicing when they are in need of repairs. Spare cylinders and valves may be necessary, especially if a machine cannot be shut down long enough to repair a component.

PREVENTIVE MAINTENANCE

Some industrial plants foster a preventive maintenance program for machines containing fluid power equipment. In these plants, the fluid power system is overhauled completely at specified intervals. The length of these intervals depends largely on the severity of the operation of the machine. Some plants service all of the machines during the vacation period; other plants may service them more often.

Many maintenance departments set up a card record system for their preventive maintenance program (Fig. 4). A record indicating the date of changing the hydraulic oil and the condition of the oil should be kept for machines that have hydraulic systems.

419

A good record system, plus a workable preventive maintenance program, aids in keeping production machines operating efficiently with minimum downtime.

REVIEW QUESTIONS

1. List five spare parts that should be kept in stock for a hydraulic system.

2. What items should be carried in stock for a rotating hydraulic cylinder?

3. What repair items have a long "shelf life?"

4. How should repair items that are subject to rust be stored?

5. Who should be contacted to determine whether the oil in a hydraulic power unit needs changing? How can he determine this?

CHAPTER 22

Servo Systems

Some of the foregoing chapters have discussed the various individual parts or components. In many instances, the performance of a combination or system (consisting of several components) is very important. For example, in an automobile each component must perform properly if the automobile is to perform satisfactorily.

CONTROL SYSTEMS

Automatic control systems have become more common. Automatic control may involve the regulation of a given function or variable, such as temperature, speed, pressure, force, displacement, or velocity, in accordance with a desired operation that can be accomplished without direct human action or attention. Control systems are used to regulate or to govern a flow of energy.

For a specific example, the thermostat in a room or building controls the heating or cooling system. A thermostat control system also may be used to start and to stop the compressor in a household refrigerator, so that a desired temperature setting can be maintained. Control systems of this type reduce the need for human attention to control the operations.

A steam turbine speed-control system is shown in Fig. 1. The steam turbine is used to drive a load, such as a generator of electrical power. Since it is desirable to maintain constant speed, the output shaft of the turbine can be regulated by adjusting the flow of steam into the turbine. The shaft speed is controlled with the aid of a flyball-type governor. The governor shaft is connected to the steam turbine shaft by means of a linkage system, such as a system of gears and levers. The flyballs or weights (with suitable links) rotate around the vertical shaft of the gov-

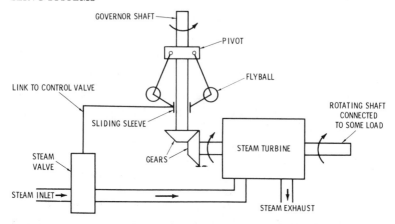

Fig. 1. Diagram of a steam turbine speed control with a flyball governor.

ernor. The vertical position of the flyballs (and the connected sliding sleeve) depends on the rotating speed of the turbine. A linkage system (gears, levers) connecting the sliding sleeve of the governor to the control valve can be used to open and to close the steam valve. If turbine speed is reduced, the flyball weights fall, admitting more steam to the turbine (by opening the control valve); thus the turbine speed is increased. In a similar manner, the turbine speed can be reduced by closing the control valve, thereby reducing the amount of steam admitted to the turbine.

The main features of the speed-control system for a steam turbine are illustrated in Fig. 1. The amount of energy involved in opening and closing the control valve is small, when compared with the output energy of the turbine shaft. The valve controls the flow of energy (the energy of the incoming steam). The flyball governor may be regarded as a device that receives a signal—a drop in speed, for example; it responds to the signal, transmitting a "command" or "input signal" to the control valve. The governor seems to be able to sense a "state" or condition, and then "feeds back" a signal to the control valve. The "signal" and "feed back" are important features in control studies.

It is possible to control a steam turbine manually, that is, without the governor system. Then, the operator must receive an indication of the actual speed of the engine, compare that speed

422

with the required speed, decide which way to rotate the valve, and then rotate the valve in the required direction. In this instance, human or manual control is required to perform the various functions of the automatic control system.

DIFFERENTIAL SENSING OR ERROR-DETECTING DEVICES

The differential sensing (difference or error-detecting) device is an important part of a control system. This device functions in the same manner that the operator performs in a manual control system.

To control the linear displacement X_o, an output rod or shaft may be connected to either a hydraulic or a pneumatic cylinder. Another shaft (called input, reference, signal, or command shaft) moves a given desired displacement X_i (Fig. 2A). For example,

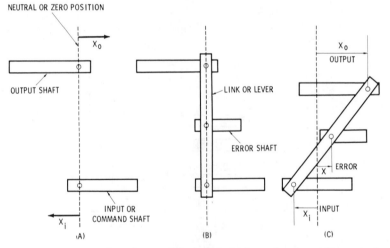

Fig. 2. Diagram of a lever differential device, showing (A) neutral position; (B) link or lever; and (C) error.

in a specific machine operation, a known variation of X_i as a function of time may be desirable. Then, the problem is to arrange for a comparison of the two displacements (X_i and X_o), so that a suitable control action can be devised.

As illustrated in Fig. 2B, the output shaft and the input shaft may be connected by a link or lever with suitable, slotted pin joints. If the input X_i and the output X_o vary as shown in the diagram, the error shaft indicates the difference X, or "error" (Fig. 2C). The error can be calculated by the formula ($X = X_o - X_i$). This lever system then serves as a differential device for linear displacements.

Error-detectors may be used to determine quantities other than the mechanical linear and angular displacements. For example, a mechanical lever system may be used to detect a difference between the input and the output forces. A differential piston arrangement (with different fluid pressures on each side of the piston) may be used to detect a pressure error signal.

A gear-differential sensing or error-detecting device is illustrated in Fig. 3. The input or command shaft A drives gear B.

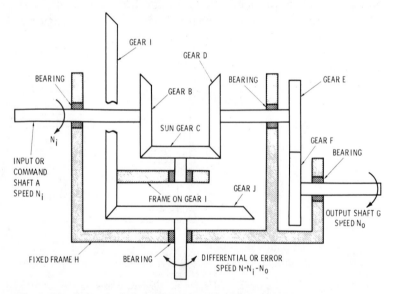

Fig. 3. Diagram of a gear-differential sensing or error-detecting device.

Gear B meshes with the sun gear C, which rotates in a frame on gear I. Gear D drives gear E which, in turn, drives gear F. Gears E and F rotate the output shaft G in the same direction that shaft A rotates. The shaft of each of the gears B, D, E, F, I, and J rotates in a bearing in the fixed frame H. The sun gear C rotates in a bearing carried on gear I. The input or command shaft A drives gear B at the desired input shaft speed N_i. If a signal from a load drives the output shaft G at speed N_o, if each gear possesses the correct number of teeth to provide a desired relation between shaft speeds, and if speed N_i equals speed N_o, neither gear I nor gear J rotates. If the input speed N_i differs from the output speed N_o, then gear J provides the differential speed $(N = N_i - N_o)$. The speed N is the "error" speed. A schematic diagram for an error-sensing device is illustrated in Fig. 4.

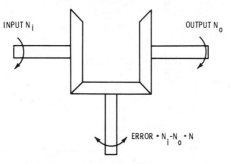

Fig. 4. Schematic diagram of an error-sensing device.

TYPES OF SERVO SYSTEMS

The hydraulic system illustrated in Fig. 5 may be studied to explain several basic concepts. Hydraulic fluid from a pump may be used to drive a rotary fluid motor. The fluid motor, in turn, drives an output load. The spool in the control valve can stop the flow of fluid; then the speed of the output load is zero. The spool of the valve can be shifted to various positions to rotate the output shaft in various directions at different values of the output speed N_o. For manual operations, an operator can actuate the control valve to control the speed of the driven output load; therefore, an error-sensing device and a mechanical connection to the control valve are not needed. A system that is controlled manually is classified as an "open-loop" system; it is not operated entirely by mechanical devices. Control is automatic, however, in Fig. 5A, and the system is classified as a "closed-loop" system; this indicates that manual control is not involved directly.

In the system (see Fig. 5A), an error-sensing device is placed at the output load. The load speed N_o is compared with the com-

425

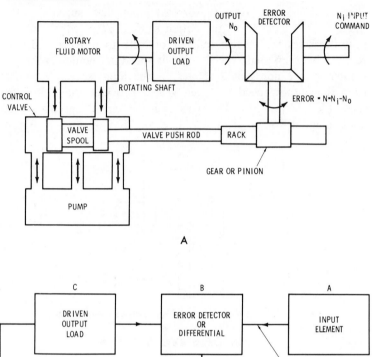

Fig. 5. Diagrams of a hydraulic servo system, showing (A) automatic control; and (B) block diagram of functional operations.

mand or desired input speed N_i by an error-detecting or differential device (a lever system or differential gear system). The differential device develops an error signal which is fed back to the control valve. If the error signal involves rotation of a shaft, the shaft rotation can be linked to the push rod of the valve by an arrangement, such as a rack and pinion. The error signal may

426

actuate the valve and change the flow of fluid to the rotary motor, thus changing the speed of the driven output load. The chief objective is to devise a control system to reduce the error.

A block diagram (see Fig. 5B) illustrates the functional operations or activities. The input element *A* develops the input or command signal which is fed into the detector *B*. The error signal from the error detector *B* is fed back to the power source and controller *D*. The controller *D* adjusts the energy flow from the power source to the driven output load *C* to correct or to reduce the error.

A "servo" or "servomechanism" system is illustrated in Fig. 5. The term "servo" is derived from the word "slave." A large variety of servo systems is possible; many different components are available for a large number of applications. The components may be hydraulic, pneumatic, or a combination of mechanical, hydraulic, and pneumatic components. The arrangement, design, and selection of components for a servo system is an art. Various service and performance requirements must be considered.

Servo systems are used to control ships, missiles, space craft, locomotives, aircraft, gun turrets, power plants, mills, presses, and various machines. A servo system may be simple; or it may be complex, depending on the nature of the functions involved and the required accuracy and response of operation.

CHARACTERISTICS OF SERVO SYSTEMS

In selecting, designing, and developing servo systems, such features as stability, speed of response, and accuracy of response must be considered, in addition to the usual cost factors—availability, reliability, and maintenance. For example, in some instances, either a hydraulic valve activated electrically or an electrohydraulic servo valve may be considered for quick response. In this type of valve, a low-level electrical input signal may result in a high-power hydraulic flow. In other instances, however, an electrohydraulic valve may not be the proper choice.

The term *stability* refers to the tendency of a body, or of some part of a servo system, to return to its original position after it has been disturbed. Both static and dynamic actions may be involved.

427

A pendulum and a pivot, an arm or link, and a mass or ball are shown in Fig. 6 to illustrate stability. In position *A*, the ball is at rest, or in static equilibrium, assuming no friction at the pivot. If the ball is displaced to position *B*, held in this position for a short time, and then released, the ball tends to return to its initial position. The ball moves beyond its initial position because of inertia. Thus, with no friction in the pivot, the ball oscillates or vibrates back and forth about the initial static equilibrium position *A*. This oscillating action is sometimes called a "hunting" or "overshooting" action. Hunting or dynamic oscillation is not desirable in a servo system. The output load does not respond properly to the error signal. If friction were introduced, however, the oscillation decreases as the period of time increases.

In Fig. 6, a pointed rod is balanced on a horizontal flat surface as another illustration of stability. A slight disturbance causes

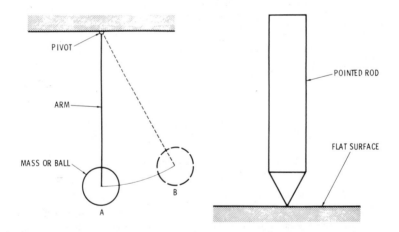

Fig. 6. Diagrams illustrating stability: the pendulum (left) illustrates "stable" static equilibrium; and the pointed rod (right) balanced on a horizontal flat surface illustrates "unstable" static equilibrium.

the rod to fall, and it does not return directly to its initial position of static equilibrium of its own accord. In this instance, the pointed rod is "unstable" in its position of static equilibrium.

A complete mathematical analysis of a complicated servo system can become much involved and very complicated. Various analytical techniques have been devised to study the stability of servo systems. In general, a differential equation is set up for the system, and the characteristics of the solution are studied to determine stability.

In reference to the diagram of a servo system (see Fig. 5B), an error signal from the error detector B is fed back to the power source controller D. The control action, in turn, determines the output C of the driven load. Various types of controllers are possible, depending on the components used and the arrangement of the components.

If an error signal N is fed into the controller D, it is possible to arrange a control or correcting action that is directly proportional to the error. The correction is small for a small error; for a larger error, the correction is larger.

It is also possible to arrange a control or correcting action that is directly proportional to the time rate of error. If the error is changing at a high rate, the correction is large; if the error is changing at a slower rate, the correction is smaller. This type of correction involves a recognition of possible error in the future. A control action that is directly proportional to the time sum or time integral of the error is possible; in this instance, the product of the error multiplied by a time change is totaled for a period of time. This type of correction involves a recognition of the history of previous error.

Various combinations of control may be arranged, such as a combination of error, error rate, and error integral. Friction also may be included. In each instance, a study should be made of the resulting characteristics, and a decision should be reached as to the combination that is most suitable for a desired service requirement.

REVIEW QUESTIONS

1. What automatic controls can be found in most homes? What causes them to operate?

2. What automatic controls can be found in most places of business? What causes them to operate?

3. List some of the characteristics of a servo system.

4. What are the effects of dirt and heat on servo systems?

5. List some of the places where a servo system might be employed.

Appendix

SUGGESTED REFERENCES

1. Stewart, Harry L. *Hydraulic and Pneumatic Power for Production*. New York: Industrial Press, 1955, 1970.

2. *Compressed Air Handbook*. New York: Compressed Air and Gas Institute, 1961.

3. Ernst, Walter. *Oil Hydraulic Power and its Industrial Applications*. New York: McGraw-Hill Book Company, 1960.

4. Hadekel, R. *Hydraulic Systems and Equipment*. New York: Cambridge University Press, 1954.

5. Pippenger, J. J. and R. M. Koff. *Fluid-Power Controls*. New York: McGraw-Hill Book Company, 1959.

6. "Aircraft Hydraulics," Naval Training Course, NavPers 10332-A. Washington, D.C.: U. S. Government Printing Office, 1951.

7. Conway, H. G. *Fluid Pressure Mechanisms*. New York: Pitman Publishing Corp., 1954.

8 McNeil, I. *Hydraulic Operation and Control of Machines*. New York: The Ronald Press Company, 1954.

9. Wilson, Warren E. *Positive-Displacement Pumps and Fluid Motors*. New York: Pitman Publishing Corp., 1950.

10. Blackburn, F., G. Reethof and J. L. Shearer. *Fluid Power Control*. New York: John Wiley & Sons, Inc., 1960.

11. "Proceedings of The National Conference on Industrial Hydraulics," Illinois Institute of Technology, Chicago, Ill.

12. *Hydraulics and Pneumatics* Magazine.

13. *Lubrication.* Texaco Technical Publication.

14. *Basic Hydraulics.* Columbus, Ohio: Division of Vocational Education, Department of Education, 1958.

15. "A Beginner's Course in Basic Pneumatics," *Applied Hydraulics*, 1954.

16. *Hydraulic Oils and Their Applications, Sun Oil Co. Technical Bulletin B-4.* Sun Oil Co., 1942.

17. Nightingale, J. M. "Hydraulic Servo Fundamentals," *Machine Design*, 1956-1958.

18. *Hydraulics As Applied To Machines.* Dearborn, Michigan: Henry Ford Trade School, 1945.

19. *Industrial Hydraulics Manual.* Detroit, Michigan: Vickers Inc.

20. *Hydraulic Controls On Machine Tools.* Flint, Michigan: General Motors Institute.

GLOSSARY OF FLUID POWER TERMS

actuator—A device that converts fluid energy to mechanical motion.

automatic controls—Those controls that are actuated in response to the cycle of the equipment.

back connected—Piping connections that are located on normally unexposed surfaces of hydraulic or pneumatic components.

channel—A fluid (either oil or air) passage that has a greater longitudinal dimension than cross-sectional dimension.

circuit—An arrangement of the component parts; or fluid equipment interconnected to a specific appliance or appliances.

clarifier—A device that removes deleterious solids and assists in maintaining the chemical stability of the hydraulic fluid.

cleaner—A device that removes solids from a fluid; the resistance of these solids to motion is a straight-line resistance.

compartment—A space within the base, frame, or column of a machine or component.

compressor—A device that converts mechanical energy to air or pneumatic energy.

compressor, fixed-displacement—A compressor that delivers a relatively constant volume of air per cycle.

compressor, variable-displacement—A compressor in which the volume of air per cycle can be varied.

cylinder—A linear motion device in which the thrust or force is proportional to the effective cross-sectional area and to the pressure change.

cylinder, single-acting—A cylinder in which the fluid (either oil or air) force can be applied only in one direction.

cylinder, plunger-type—A cylinder in which the internal element is constructed with a single diameter and a contracting type of seal.

cylinder, piston-type—A cylinder in which the internal element is designed with one or more diameters and an expanding type of seal.

cylinder, double-acting—A cylinder in which the fluid force can be applied in either direction.

cylinder, single-end rod—A cylinder designed with the piston rod extending from one end of the cylinder.

cylinder, double-end rod—A cylinder designed with two piston rods—one piston rod extending from each end of the cylinder.

enclosure—A housing designed for hydraulic or pneumatic apparatus.

filter—A device that is used to remove solids from a fluid; the resistance to motion of these solids is in the form of a tortuous path.

433

front connected—Piping connections that are normally placed on the exposed surfaces of hydraulic or pneumatic components.

hydraulic panel, mounting—A plate on which hydraulic components can be mounted.

hydraulic panel, control—Grouping of hydraulic controls in units to form a single assembly either on a mounting plate or inside a casting; a single mounting surface may be used for the entire unit.

line—A tube, pipe, or hose that acts as a conductor of hydraulic fluid or air.

line, drain—A line that independently returns excess or leakage oil to the reservoir or vented manifold.

line, hydraulic exhaust—A return line which carries power or control actuating hydraulic fluid back to the reservoir.

line, pneumatic exhaust—A return line which carries the power or control actuating air back to the atmosphere.

line, joining—A hydraulic or pneumatic line that either crosses or connects with another line either on a schematic diagram or in actual construction.

line, passing—A hydraulic or pneumatic line that crosses another line on a schematic diagram, but it does not connect with another line in actual construction.

line, pilot—A line that conducts a control hydraulic fluid or air.

line, working—A hydraulic or pneumatic line that acts as a conductor of power-actuating hydraulic fluid or air.

lubricator—A device that is used to add lubricant to the actuating air.

manual controls—Those controls that are actuated by the operator, regardless of means.

mass production—A method in which a setup is used to produce a number of identical workpieces for an indefinite period of time.

motors and cylinders—Devices that convert hydraulic or air energy into mechanical energy.

motor, oscillating—A motor that produces a maximum angular rotating movement of less than 360 degrees in either direction.

motor, rotary—A motor that produces rotary motion; the torque output is proportional to the fluid displacement per revolution and to the pressure change between intake and exhaust ports.

motor, rotary, fixed-displacement—A rotary motor in which the displacement per revolution cannot be adjusted.

motor, rotary, variable-displacement—A rotary motor in which the displacement per revolution can be adjusted.

muffler—A device that is used to reduce or muffle exhaust noises.

passage, hydraulic—A machined or cored connection that conducts hydraulic fluid within or through a hydraulic component.

passage, pneumatic—A machined or cored connection that conducts air within or through a pneumatic component.

phase of cycle—(0) Neutral; (1) Rapid Advance; (2) Feed or Pressure stroke (Forward and Return); (3) Dwell; (4) Rapid Return.
1. *Rapid Advance*—The approach of the tools or workpiece to the feed position.
2. *Feed*—The portion of the cycle where work is performed on the workpiece.
3. *Dwell*—The portion of the cycle where the feed rate or pressure stroke is stopped.
4. *Rapid Return*—The return of the tools or workpiece to the cycle starting position.

pneumatic mounting panel—A plate on which pneumatic actuating components are mounted.

pneumatic control panel—A grouping of pneumatic control units having a single mounting surface, and mounted as a single assembly on a plate or inside a casting.

port—An internal or external opening on the surface of a component or at the end of a passage.

port, valve—A controlled opening between passages—an opening that can be closed, opened, or modulated.

positive position stop—A structural member that limits the motion of the work at a desired position.

positive safety stop—A fixed structural member which limits maximum travel to the design limits of the machine or equipment.

pressure seals—Seals that increase their sealing action as fluid pressure increases.

pump—A device that converts mechanical energy to hydraulic energy.

pump, fixed-displacement—A pump which delivers a relatively constant volume of hydraulic fluid per cycle.

pump, variable-displacement—A pump in which the volume of hydraulic fluid per cycle can be varied.

restriction—A device which causes a deliberate pressure drop or resistance in a line or passage by means of a reduction in cross-sectional area.
   Choke—A restriction with a longitudinal dimension that is larger than its cross-sectional dimension.
   Orifice—A restriction with a longitudinal dimension that is smaller than its cross-sectional dimension.

restriction choke fixed viscous—A nonadjustable resistor in which the pressure drop, or rate of flow, is not entirely independent of viscosity.

restriction orifice (fixed nonviscous)—A nonadjustable resistor in which the pressure drop, or rate of flow, is relatively independent of viscosity.

schematic diagram—A drawing or drawings indicating the functional construction of valves, controls, and actuating mechanisms.

sealing device—A part of an assembly or parts that are used to prevent leakage between two or more parts.

separator—A device that is used to separate water or other materials that have specific weights that are different from the actuating medium.

strainer—A device that is used to remove solids from a fluid; the resistance to motion of these solids is a straight-line resistance.

subplate (back plate)—An auxiliary mounting plate on which piping connections are mounted for a hydraulic or pneumatic actuating component.

surge—A transient rise of hydraulic or air pressure in the fluid circuit.

trip device—A mechanical element that is used to actuate a position control.

unit production—A method which provides a single workpiece per setup.

CONVERSION FACTORS

Measures

1 foot = 12 inches
1 square foot = 144 square inches
1 cubic foot =1728 cubic inches
1 cubic foot = 7.48 gallon
1 inch = 25.4 millimeters
1 inch = 2.54 centimeters
1 millimeter = 0.03937 inch
1 meter = 39.37 inches
1 micron = 0.000001 meter
1 gallon = 4 quarts
1 quart = 2 pints
1 gallon = 231 cubic inches
1 Imperial gallon = 1.2009 gallons
1 gallon = 0.833 Imperial gallon
1 cubic foot of water weighs approximately 62.4 pounds (at 60 degrees F.)
1 liter = 2.113 pints
1 gallon = 3.785 liters

1 kilogram = 1000 grams
1 kilogram = 2.205 pounds

Pressure

1 standard atmosphere = 14.7 pounds per square inch, absolute
1 standard atmosphere = 29.92 inches of mercury
1 standard atmosphere = 33.4 feet of water (at 60 degrees F.)
1 inch of water (at 60 degrees F.) = 0.0361 pounds per square inch

Rate of Motion

1 gallon per minute = 3.85 cubic inches per second
1 gallon per minute = 0.002228 cubic feet per second
1 foot per second = 0.3048 meter per second
1 meter per second = 3.2808 feet per second

Power And Work

1 horsepower = 550 foot-pounds per second
1 horsepower = 33,000 foot-pounds per minute
1 horsepower = 745.7 watts
1 horsepower = 2545 Btu per hour
1 Btu = 778 foot-pounds
1 kilowatt = 1000 watts
1 watt = 44.26 foot-pounds per minute

Prefixes

micro- = one-millionth
milli- = one-thousandth
centi- = one-hundredth
deci- = one-tenth
kilo- = one thousand
mega- = one million

Chapter 1

ANSWERS TO REVIEW QUESTIONS

1. Hydraulic brakes are used to bring the automobile to a stop. They are much smoother in operation than the earlier mechanical brakes, and require less maintenance.

2. Pneumatic devices do not present the hazard of electrical shock which can occur with an electrical device, and they are especially useful in hazardous conditions where gas or other volatile materials might be present.

3. The relief valve used in conjunction with a hot-water heater protects the heater from excessive build-up of pressure which can cause an explosion.

4. Pneumatic devices are used for road and street repairs, because they:
   a. In many instances, electricity is not available.
   b. Are quite portable.
   c. Are rugged and built for outdoor use.
   d. Are designed to produce a wide range of forces.
   e. Can be used in adverse conditions, as in water, hazardous locations, etc.

5. The increased requirements or demands to automate work processes to meet competition have resulted in increased use of fluid power devices, due to their reliability, simplicity, low maintenance, and versatility.

6. Examples of uses for fluid power equipment in processing food are:
   a. To provide the motions for casers and un-casers in the food and beverage industry.
   b. In labeling cans and bottles, much pneumatic equipment is employed to provide the various motions.
   c. Hydraulic-operated lift trucks move pallets of boxed can goods to various locations in warehouses and to transport trucks.

d. Hydraulic-operated presses are used to form and press many of the meat and fish products.

e. In many instances, bakery equipment, such as ovens, mixers, blenders, sifters, etc., are operated by fluid power devices.

Chapter 2

ANSWERS TO REVIEW QUESTIONS

1. The oil gauge on the automobile dash visually indicates the oil pressure.

2. The work done is:

   20 feet $\times$ 170 pounds = 3400 foot-pounds

3. The horsepower expended is:

$$\frac{3400 \text{ ft.-lb.}}{20 \text{ seconds}} = \frac{170 \text{ ft. lb.}}{\text{seconds}}$$

$$1 \text{ horsepower} = \frac{550 \text{ ft.-lb.}}{\text{seconds}}$$

$$\frac{170}{550} = 0.309 \text{ horsepower}$$

4. Oil pressure needed is:

$$\frac{2000 \text{ lb.}}{4 \text{ sq. in.}} = 500 \text{ lb. per sq. in.}$$

5. The capacity, in gallons per minute, that the pump must deliver is:

   4 sq. in. $\times$ 60 in. = 240 cu. in. in 1 second

   231 cu. in. in 1 gallon

   240 cu. in. $\times$ 60 seconds = 14400 cu. in. per minute

$$\frac{14400}{231} = 62.2 \text{ gallons per minute}$$

6. A torque wrench can be set to tighten a screw or bolt a set amount. When this setting is reached, the wrench will not apply

any more force on the screw or bolt. This prevents the threads from being stripped or the screw or bolt from being twisted off.

7. Pressure is the force per unit area, for example, pounds per square inch; whereas, force is the resultant of pressure times area.

8. Work is the product of force times displacement, as in moving an object from A to B; whereas, torque is used in connection with turning and twisting actions. Torque can be defined as the product, FS, where F is the tangential force and S is the radius.

Chapter 3

ANSWERS TO REVIEW QUESTIONS

1. The common bicycle pump can be classified a fixed-displacement machine, provided the full stroke of the pump is always used.

2. Advantages of a fixed-displacement pump are:
   a. Generally simple in construction.
   b. Relatively low in cost.
   c. Compact in size.
   d. Readily available.

3. Advantages of a variable-displacement fluid motor are:
   a. Peak performance at peak requirements.
   b. Reduction in horsepower requirements.
   c. Choice of actuators to control the variable-displacement mechanism.
   d. Compact in size.

4. Reciprocating fluid motors can be used to:
   a. Actuate a control valve, such as a butterfly valve, which can be used to control large volumes of water.
   b. Stroke a large press.
   c. Actuate a turnover device on a conveyor.
   d. Actuate the brakes on a car.
   e. Actuate clamping devices.

5. Rotary fluid motors can be used to:
    a. Operate machine spindles.
    b. Actuate reels or drums of winding devices.
    c. Operate rotary mixing devices.
    d. Operate power steering devices.
    e. Operate slides on machines through a gear train.

6. A compressor is used to compress air or a gas; a pump is used used to move a liquid under pressure.

7. A pump can be protected from overloading by using a relief valve in conjunction with the pump. The relief valve relieves the pressure when it exceeds the pressure setting of the relief valve.

Chapter 4

ANSWERS TO REVIEW QUESTIONS

1. Drawings using *ANS* Graphic Symbols are:

    a.

    b.

2. Sketched diagrams, using *ANS* Graphic Symbols are:

    a.

b.

c.

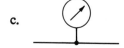

d.

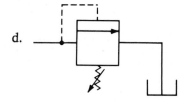

e.

3. Sketched diagram, using *ANS* Graphic Symbols.

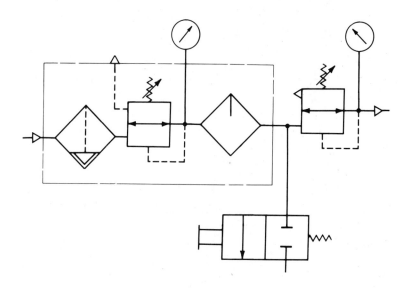

4. Diagram of Fig. 1, Chapter 1, using *ANS* Graphic Symbols.

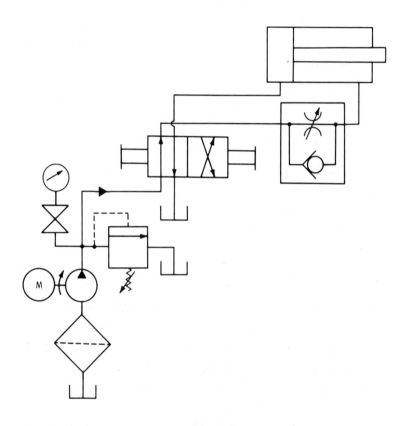

5. Diagram of Fig. 2, Chapter 1, using *ANS* Graphic Symbols.

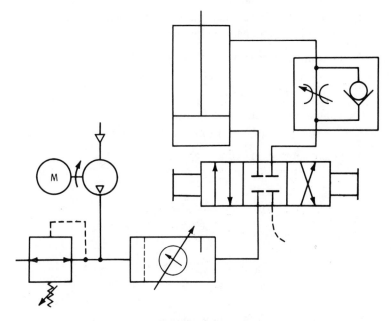

Chapter 5

ANSWERS TO REVIEW QUESTIONS

1. A pressure booster is a device used to increase the secondary pressure of the system above that of the primary pressure. If the primary pressure is 100 *psi,* the secondary pressure might be 2000 *psi.* This would be a ratio of 20 to 1.

2. Advantages of a pressure booster are:
    a. High output in *psi.*
    b. Heat does not become a problem.
    c. High-pressure pumping system is not needed.
    d. Various fluids can be used in the primary system.

3. Possible causes of failure of pressure boosters are:
    a. Leakage around the seals in booster chamber.
    b. Booster does not have capacity to complete the work cycle.

    c. Dirt in primary system can cause damage to the precision parts of the booster.

    d. In a pneumatic-actuated booster, lack of lubrication is a cause of seal and packing failures.

4. The intensified pressure is:

$$\frac{12.57 \text{ sq. in.} \times 1500 \text{ psi}}{0.7850 \text{ sq. in.}} = 24000 \text{ psi}$$

5. The volume of the intensified fluid is:

$$0.785 \text{ sq. in.} \times 12 \text{ in.} = 9.42 \text{ cu. in.}$$

6. Diameter required on cylinder end is:

$$25000 \text{ psi} \times 1.25 = 750 \times X$$
$$X = \frac{31250}{750}$$
$$= 41.66 \text{ sq. in.}$$

$$\text{Dia. of piston} = 7.28 \text{ in.}$$
$$1.5 \text{ in. Dia. ram area} = 1.767 \text{ sq. in.}$$

7. The volume of intensified fluid delivered per stroke is:

$$1.767 \text{ sq. in.} \times 20 \text{ in. stroke} = 35.34 \text{ cu. in.}$$

8. The intensified pressure is:

$$113.1 \text{ sq. in.} \times 100 \text{ psi} = 1.767 \text{ sq. in.} \times P$$
$$P = \frac{113.3 \text{ sq. in.} \times 100 \text{ psi}}{1.767 \text{ sq. in.}}$$
$$P = 6400 \text{ psi}$$

9. Advantages of using water, rather than oil, for the intensified medium are:

    a. Water is more easily removed from the pipe after testing.

    b. Water is less expensive.

448

10. The advantage of using an intensified fluid, rather than a mechanical device, is that an intensified fluid will exert equal force in all directions, as stated in Pascal's law.

11. Sketches of:

    a. Single-acting booster actuated by air to intensify water pressure. Piston returned by spring pressure.

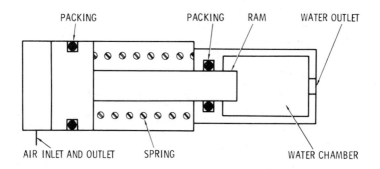

    b. Double-acting booster, intensification in one direction, actuated by oil for intensifying oil pressure.

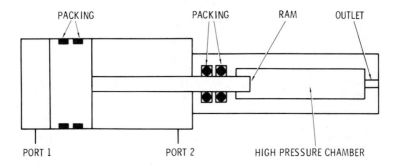

12. Sketch showing proper connection of booster to a hydraulic cylinder.

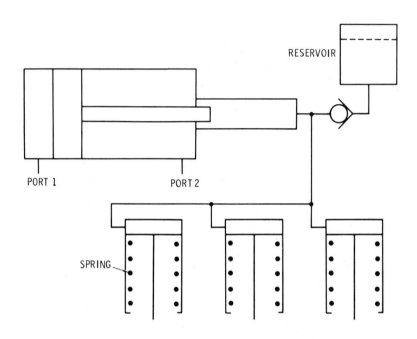

RESERVOIR

PORT 1   PORT 2

SPRING

Chapter 6

ANSWERS TO REVIEW QUESTIONS

1. The foundation for an air compressor should be solid to carry the weight of the compressor. Vibration from the compressor must be absorbed by the foundtion.

2. Lower temperatures and moisture content improve the operating conditions of the compressor. Moisture is especially harmful to a pneumatic system.

3. The air receiver serves as a storage tank for the compressor and stores compressed air to meet peak demands. The air

receiver also serves as a dampening device for a recipro-cating-type compressor.

4. The aftercooler is used to cool the compressed air as it flows from the compressor; the cooling separates the moisture from the compressed air. The moisture is then drained off by a water separator.

5. The safety valve should be located at the receiver, unless there is no stop valve between the compressor and the re-ceiver. If there is a stop valve between the compressor and receiver, the safety valve should be located between the compressor and the stop valve.

6. Since it has moving parts, proper lubrication is important for the compressor. The crankcase may require one type of lubricant, the cylinder another type, and the bearings still another type of lubricant.

7. In a single-stage compressor, the air is compressed once be-fore it moves to the discharge port; whereas, in a two-stage compressor, the air is compressed twice before it moves to the discharge port.

8. Schematic diagram of Fig. 12, using *ANS* Graphic Symbols.

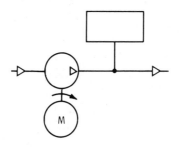

9. Schematic diagram of Fig. 14, using *ANS* Graphic Symbols.

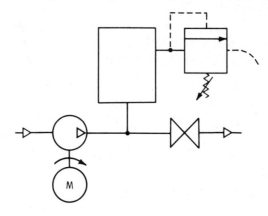

10. Schematic diagram of Fig. 15, using *ANS* Graphic Symbols.

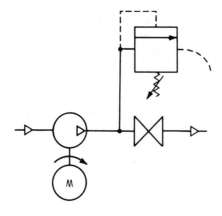

Chapter 7

ANSWERS TO REVIEW QUESTIONS

1. A hydraulic power device is the power source in a hydraulic system.

2. Advantages of gear-type pumps are:
   a. Low in cost.

b. Compact.

c. Relatively troublefree in operation.

3. A "hi-low" pumping system is one which delivers a large volume of liquid at low pressure and a small volume of liquid at high pressure. The changeover from one condition to the other is automatic.

4. The reservoir:
   a. Serves as a storage tank for the liquid.
   b. Serves as a means of cooling the liquid.
   c. Allows impurities to settle and be pumped from the system.
   d. Serves as a base for mounting the pump, electric motor, and some controls.

5. The baffles in a reservoir separate the oil that is returned from the system from the oil that is ready to be pumped into the system. Baffles break up the bubbles and foam in the return oil; also, they cause the impurities in the return oil to drop to the bottom of the reservoir.

6. The intake filter prevents the larger impurities or dirt in the oil from moving through the pump.

7. The breather assembly is used on a reservoir to insure that atmospheric pressure contacts the surface of the oil in the reservoir.

8. The causes of a pump becoming noisy are:
   a. Insufficient oil in reservoir.
   b. Clogged intake filter.
   c. Improper oil viscosity.
   d. Misalignment between pump and electric motor.
   e. Leaky shaft seal.
   f. Worn parts, such as bearings, vanes, etc.
   g. Cavitation.

9. An abnormally low rate of flow from a pump is caused by:
   a. Cavitation.
   b. Incorrect electric motor speed.
   c. Relief valve malfunction.

    d. Worn parts in pump.

    e. Improper oil viscosity.

10. An abnormally low pump pressure is caused by:

    a. Malfunction of relief valve.

    b. Leak in inlet line to pump.

    c. Worn parts in pump.

    d. Worn parts in control valves or cylinders.

    e. Porosity in pump body.

    f. Open valve in system.

11. A pump can be caused to stall by:

    a. Malfunction of relief valve.

    b. Improper electric motor.

    c. Frozen bearings in the pump.

    d. Broken internal parts of the pump.

12. A hydraulic power device should be mounted:

    a. In a clean well-ventilated location where the ambient temperature is not abnormal.

    b. On a good solid foundation.

    c. At the lowest point in the hydraulic system so that drain lines have unrestricted flow.

    d. In a readily accessible location so that it can be serviced.

13. A heat exchanger in a hydraulic system keeps the oil temperature at a range that seals will not become hard and brittle; keeps the oil viscosity at a satisfactory range to provide proper lubrication; keeps contraction and expansion at a minimum in the hydraulic components which, in turn, would control the internal leakage and allow efficient operation of the system.

14. Heaters are used in cold locations, such as cold storage rooms in packing plants; in cold outdoor operations; or in buildings where temperatures may drop to the freezing point or below, as in steel mills or other processing mills.

Chapter 8

ANSWERS TO REVIEW QUESTIONS

1. Three disadvantages of using water in a fluid power system are:
    a. Water can freeze in a system located in cold climates.
    b. Water is corrosive and can cause trouble to precision parts of pumps and controls.
    c. Water can cause "wire drawing" in components which, in turn, causes internal leakage.

2. Causes of hydraulic oil becoming overheated are:
    a. Oil spilling over the relief valves at high pressure.
    b. High ambient temperatures around the system.
    c. Poor ventilation.

3. Three ways in which air can enter a hydraulic system are:
    a. Through leaks in the intake to the pump.
    b. Oil level that is below the intake filter part of the time.
    c. Leak in the pump seal or in other seals in the system.

4. Dirt can enter a hydraulic system:
    a. Through the use of certain oils; excessive heat in the system often causes varnish deposits.
    b. An air breather on the reservoir might be removed.
    c. Poor installation in which pipe scale, pipe compound, or metal chips from pipe threads enter the system.
    d. Sand may break away in cored passages under pressure, in cart valve, or pump bodies.
    e. Moisture in the oil can cause rust, if the system remains inoperative for a period of time.

5. Precautions in changing oil in a hydraulic system are:
    a. Make certain the oil is clean and free of lint.
    b. Keep moisture out of the oil.
    c. Make sure the reservoir and filter are clean.
    d. Use oil suitable for the system.

6. Precautions in storing hydraulic oil are:
    a. Keep oil in a clean container.

b. Keep cover on container tight.

c. Keep oil in a dry location.

7. Hazards presented by hydraulic oil on the floor are:
a. Causes slipping condition.
b. Presents a fire hazard.

8. Three types of commonly used hydraulic oil are:
a. Petroleum-base liquid.
b. Synthetic-base mixtures.
c. Water-base fluids.

9. Effects of fire-resistant fluids on packing, gaskets, filters, etc. are:
a. Causes some packings to swell.
b. Shortens packing life.
c. Attacks some metals in filters.
d. Attacks some paints used in reservoirs or in valve bodies.

Chapter 9

ANSWERS TO REVIEW QUESTIONS

1. Classes of piping in a home are:
a. Rigid—water lines and gas lines.
b. Semirigid—water lines and gas lines.
c. Flexible—on an automatic washer.

2. Classes of piping found in an automobile are:
a. Rigid—miscellaneous uses.
b. Semirigid—lines to brakes, to gas tanks, etc.
c. Flexible—water hoses, heater hoses, wiper hoses, etc.

3. Classes of piping in a workplace are:
a. Rigid—water lines and compressed air lines.
b. Semirigid—water lines and gas lines.
c. Flexible—to air tools.

4. Problems in tubing installations avoided by rigid piping are:
a. When copper tubing is used in a hydraulic system it may work-harden and fail.

b. Large pipe fittings and pipe are usually more readily available than large fittings for tubing.

c. Flaring of some tubing becomes difficult and may cause problems, such as splitting.

5. Hazards of flexible hose are:
   a. Some fluids affect flexible hose.
   b. Flexible hose may whip under high pressure, becoming dangerous to personnel.
   c. Flexible hose may be broken, cut, or flattened by a shop truck, which causes a sudden stoppage of liquid.

6. Causes of low-pressure air are:
   a. Insufficient compressor capacity.
   b. Faulty regulating valve in system.
   c. Leaks in the various air components in system.
   d. Inadequate piping.
   e. Open valve in system.
   f. Clogged filters.

7. Factors that should be checked before installation of piping are:
   a. Decide maximum required pressure, then install suitable piping.
   b. Check peak demands and use piping that will meet these demands.
   c. Design piping system to eliminate as many bends as possible.
   d. Eliminate low spots in piping layout.

8. Keeps the compound or tape from entering the system when a joint is made.

9. Problems caused by using a hacksaw are:
   a. May not cut straight.
   b. May cause small chips to enter interior of tubing.
   c. May slightly flatten thin walled tubing to a point where it is difficult to make a suitable flare.

10. Rigid piping should be used. This eliminates the possibility of collapse of the pipe; also it provides for full flow.

11. Piping should be used as follows:

a. Semirigid, such as aluminum or copper, depending on whether it is to be used for pneumatic or hydraulic service. Flexible, or hose, is satisfactory for many applications.

b. Semirigid or rigid for 2000 *psi* hydraulic service. Annealed steel tubing is very popular for hydraulic systems. Steel pipe is also used. Flexible hose of the wire braid variety is employed for many hydraulic applications.

Chapter 10

ANSWERS TO REVIEW QUESTIONS

1. Disadvantages of by-passing the filter are that it permits:
   a. Impurities, such as dirt, pipe scale, lime deposits, and rust to enter pneumatic systems.
   b. Moisture to enter the components of the system.

2. Cyclonic filtration is accomplished by mechanical means. Edge filtration is accomplished by the air passing through a layer of ribbon made of a synthetic material.

3. The pressure regulator provides a constant, set pressure at the outlet of the regulator and can be regulated to provide the optimum operating pressure for the system.

4. Reasons for placing a lubricator in the air line are:
   a. Stores oil.
   b. Sends regulated portions of oil into the system to lubricate controls, cylinders, and air motors.

5. A pressure gauge is used to visually indicate the operating pressure of the system.

6. Disadvantages of a single air compressor for a large number of air tools are:
   a. If the compressor should fail, the entire plant is without compressed air.
   b. The necessary location of the compressor may fail to provide adequate compressed air supply to the entire plant.

c. With only a single compressor in the plant, long runs of piping may become necessary, which reduces the efficiency.

7. Schematic diagram, using *ANS* Graphic Symbols.

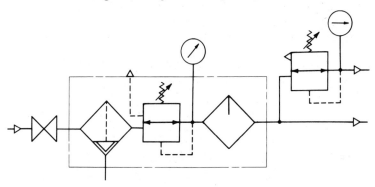

Chapter 11

ANSWERS TO REVIEW QUESTIONS

1. A three-way valve has three ports: (1) inlet; (2) cylinder; and (3) exhaust.

2. A four-way valve has one more port, which is a second cylinder port, than a three-way valve.

3. A directional control valve directs the fluid through the system. It can be a three-way, four-way, or five-way valve.

4. In a two-position four-way valve, the spool or flow director has two positions. In one position of the spool, the inlet is connected to No. 1 cylinder port and No. 2 cylinder port is connected to exhaust. In the other position of the spool, the inlet is connected to the No. 2 cylinder port and No. 1 cylinder port is connected to exhaust.

5. A manually operated valve is one in which the actuator is operated by hand or foot; whereas, in a mechanically operated valve, the actuator is operated by a cam on a machine, a trip on a press, etc.

6. A pilot-operated valve may be a two-, three-, or four-way valve, and is operated by the fluid acting on the valve actuator. It can be a double-pilot, in which the valve would have two actuators; or it can be a single-pilot, in which the actuator operates in one direction and a spring returns the spool to the starting position.

7. In a solenoid valve, the actuator is a solenoid. There are single-solenoid and double-solenoid actuators.

8. Factors causing a valve to fail are:
   a. Dirt.
   b. Improper application—pressure or volume.
   c. Excessive heat.
   d. Use of incorrect actuator; requirements of application may be too severe for actuator.

9. Size (pipe size) of valves used to pass the following volumes of oil are:
   a. 3 gpm— ⅜ in.
   b. 8 gpm— ½ in.
   c. 20 gpm—1 in.
   d. 35 gpm—1¼ in.
   e. 75 gpm—2 in.
   f. 100 gpm—2½ in.
   g. 150 gpm—2½ in.
   h. 250 gpm—3 in.

10. *ANS* Graphic Symbols for the following pneumatic valves are:

    a. Solenoid-operated, normally-open two-way valve

    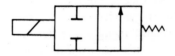

    b. Manually operated three-way valve

c. Three-position, spring-centered, closed-center, solenoid-operated four-way valve.

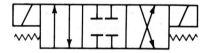

d. Air-piloted, two-position four-way valve.

11. *ANS* Graphic Symbols for the following hydraulic control valves are:

a. Two-position, pilot-operated, solenoid-operated four-way valve.

b. Three-position, closed-center, manually operated four-way valve.

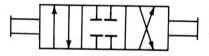

Chapter 12

ANSWERS TO REVIEW QUESTIONS

1. A flow control valve limits the volume of fluid that can flow through the valve. The volume is controlled by an orifice which can be changed in size by a simple adjustment.

2. Examples of flow control valves commonly found in a home are:
   a. A water faucet.
   b. The control on a gas burner.
   c. The gas control on a hot-water heater.
   d. A hot-air register control.

3. A metering-in flow control meters the fluid entering a system. This is usually found in hydraulic circuits.

4. A metering-out flow control meters the fluid leaving the device that is is to control. For example: if a cylinder is to be controlled on the forward stroke of the piston, the flow control meters the fluid leaving the rod end of the cylinder. The flow control is placed in the line between the rod end of the cylinder and the directional control valve.

5. A check valve in a flow control provides a checking action in one direction of fluid flow; so therefore, the fluid must pass through the orifice in the control. In the opposite direction of flow, the fluid can pass freely through the flow control.

6. The spring in a flow control keeps the check closed. Sometimes, a flow control might be installed upside down; therefore, if there were no spring, the check would remain open.

7. Factors to consider in installing a flow control are:
   a. Install the control as near as possible to the component to be controlled.
   b. Select the proper function the control will play in the system—meter-in, meter-out, etc.
   c. Install a flow control with the proper pressure rating.
   d. Install a control of sufficient capacity to do the job properly.

8. Possible causes of failure in a flow control valve are:
   a. Dirt in system.
   b. Inadequate capacity—size and pressure rating.
   c. Broken spring.
   d. Flow control installed backward in piping.

9. Examples of flow controls on automobiles are:
   a. Accelerator.
   b. Heater controls.
   c. Air conditioning.

10. Installation procedure for a cam-operated hydraulic flow control valve are:
    a. Mount the valve on a firm mounting pad, since these valves usually require a sizeable force to depress the cam roller.
    b. Mount the valve where it will not be subject to considerable dirt.
    c. Keep the valve removed from excessive heat.
    d. Mount the valve in a place that the adjustment is readily accessible.
    e. Connect the "OUT" port of the control to a cylinder port in the directional control valve; and connect the "IN" port to the rod-end port of the cylinder, if the control is to be on forward stroke of the cylinder.
    f. Grease the cam roller on the actuator.

11. A cam-operated flow control in a hydraulic system can be used:
    a. In place of cushions in a cylinder. It can provide much longer cushioning than could be economically feasible in a cylinder on high-shock applications.
    b. To provide a skip-feed arrangement.
    c. To provide a free-flow return.

Chapter 13

ANSWERS TO REVIEW QUESTIONS

1. The relief valve is designed to be a safety valve; and in a hydraulic system, it protects the pump and other system components from being overloaded. The relief valve is adjustable for various pressure settings, such as 0 to 1500 *psi*. When the set pressure is reached, fluid is released to the reservoir.

2. The pressure reducing valve is placed in the system to permit that part of the system to operate at a lesser pressure than the main part of the system.

3. The sequence valve sets up a sequence of movement, without the need for a second directional control valve. For example, when the piston in cylinder 1 reaches the end of its travel and pressure builds up, the sequence valve opens and permits fluid to flow to cylinder 2 to complete its travel in one direction. Two sequence valves, a directional control valve, and two cylinders can be used in a system for operating two cylinders in sequence, both on the outstroke and on the return stroke.

4. In a press circuit, that utilizes a high-low system, the unloading valve is used to unload, automatically, the large-volume, low-pressure pump; this keeps the oil temperature at a satisfactory level. Unloading valves are pilot-operated, and they are a type of relief valve.

5. Two relief valves are often used in press circuits. One relief valve (high-pressure) is set for high pressure; it controls the downward force of the press ram. The second relief valve (low-pressure) is used to control the pressure on the return stroke of the press ram and for the standby period.

   The second relief valve is set to provide only enough pressure to return the ram. Overheating of the liquid is reduced during the standby period.

6. Factors to consider when installing a relief valve are:
   a. Use a relief valve with sufficient capacity to amply relieve the pump.
   b. Do not restrict the exhaust line from the relief valve to the reservoir.
   c. If the relief valve is subplate-mounted, make certain that "O" ring gaskets are in place.
   d. If pipe line mounted, make certain that pipe connections are secure.
   e. Mount relief valve as near the pump as possible.

464

7. Conditions causing a relief valve to fail are:
   a. Inadequate size for the application.
   b. Broken springs.
   c. Stuck spool due to varnish deposits or dirt.
   d. Back pressure on the exhaust port.
   e. Air in system.
   f. Heat.

8. In the direct-acting valve, the main spool operates on a large spring. In the direct-operated pilot-type valve, a control head operates the main spool which has only a small spring operating against it.

9. Conditions that can cause a pressure reducing valve to fail are:
   a. Dirt or varnish deposits can cause the spool to become stuck.
   b. Dirt may clog small orifices.
   c. Broken spring.
   d. Misapplication—pressure or flow capacity.
   e. Heat.
   f. Leaky seals.

10. Conditions that can cause an unloading valve to fail are:
    a. Dirt.
    b. Broken spring
    c. Excessive shock.
    d. Back pressure.
    e. Misapplication.

Chapter 14

ANSWERS TO REVIEW QUESTIONS

1. A single-acting cylinder produces force in only one direction; whereas, a double-acting cylinder produces force in both directions of the piston.

2. A ram-type cylinder is usually single-acting, and the ram has approximately the same area as the cylinder. The ram-type

cylinder may not have packing at the rear end of the ram. A piston-type cylinder has a piston and piston rod, and the piston has some type of seal. This type of cylinder is usually double-acting.

3. A nonrotating cylinder can be used for a press cylinder, where a great amount of force is exerted on the down stroke and a lesser amount of force on the return stroke.

4. A rotating cylinder can be used in conjunction with a power chuck or other revolving holding device. The rotating cylinder has a stationary distributor.

5. A rod wiper is a synthetic device which wipes the piston rod to prevent dirt being drawn into the packing. It is usually installed in the packing gland cap. A rod scraper is a metal device and is used similarly to a rod wiper.

6. A cushion collar and hose reduces shock as the cylinder piston contacts the cylinder covers.

7. The cushion adjustment valve meters the fluid that is trapped when the cushion nose closes the main orifice. The adjustment covers a fairly wide range to compensate for various operating conditions.

8. Advantages of a double-end cylinder are:
   a. Provides extra bearing support for the piston rod.
   b. Produces the same force in both directions of stroke.
   c. One end of the piston rod can be used to operate limit switches, etc.

9. Factors causing a nonrotating cylinder to fail are:
   a. Dirt.
   b. Heat.
   c. Side thrust.
   d. Internal leakage.
   e. Broken internal parts.

10. Factors causing a rotating cylinder to fail are:
    a. Leaky distributor seals.
    b. Scored cylinder tube.

c. Excessive speed of rotation.

d. Lack of lubrication.

e. Heat.

f. Dirt.

11. Five types of rod packings used in pneumatic or hydraulic cylinders are:

a. Chevron type.

b. "O" ring.

c. "Quad" ring.

d. Block-vee.

e. "Hat" packing.

12. Five types of seals used on hydraulic cylinder pistons are:

a. "O" ring

b. "Quad" ring.

c. Automotive-type rings.

d. Block-vee.

e. Cup packing.

13. Cup packing, block-vee, or other synthetic seals are recommended for leakproof hydraulic cylinders. These seals cling to the cylinder walls to provide a tight seal.

14. Three types of seals that can be used on pneumatic cylinder pistons are:

a. Cup packing.

b. Block-vee.

c. "O" ring.

15. A boot should be used to cover a piston rod of a pneumatic cylinder to protect the precision finish of the piston rod. One end of the boot should be attached to the packing gland of the cylinder and the opposite end to the end of the piston rod. The end of the piston rod is usually furnished with a collar to fasten the boot.

16. A sheet metal cover is often installed over a piston rod to protect it from falling objects, weld spatter, etc.

Chapter 15

ANSWERS TO REVIEW QUESTIONS

1. Reasons for using pneumatic tools in mining are:
   a. No electric motors present.
   b. Eliminate possibilities of shock.
   c. Air motors can be stalled without damage.
   d. Easily portable.

2. Pneumatic drills are preferable to electric drills in lightweight assembly jobs because:
   a. Light in weight.
   b. Can usually be stalled without damage.
   c. Easily maintained.
   d. No possibility of shock.
   e. Infinite range of speeds.

3. Advantages of portable air tools for street construction and repair work are:
   a. Easily portable.
   b. Can be used in all types of weather.
   c. Rugged in construction.
   d. Wide range of forces.

Chapter 16

ANSWERS TO REVIEW QUESTIONS

1. Advantages of displacement-type hydraulic transmissions for moving gun turrets are.
   a. Compact in size.
   b. Deliver high torque when required.
   c. High degree of accuracy.
   d. Relatively maintenance free.
   e. Oil is delivered from pump to motor only when needed.
   f. Rotation of output shaft can be reversed quickly without problems.

2. This question can be debated. At present, these transmissions have not been employed, likely due to the cost. However, these transmissions are being used on expensive mobile units of various types.

3. Hydraulic motors are much smaller.

4. Four applications for fluid motors are:
   a. Machine tools—used to operate lead screws for traversing tables, etc.
   b. Hydraulic winches—supplies the torque for operating the reel.
   c. Drive wheels—supplies power for drive wheels on large earth movers.
   d. Traversing mechanisms on military equipment—supplies power to drive heavy traversing mechanisms.

5. Rough sketch, using *ANS* Graphic Symbols, of simple circuit incorporating: hydraulic power device with constant-displacement pump; four-way two-piston manually operated valve; and a fixed-displacement hydraulic motor.

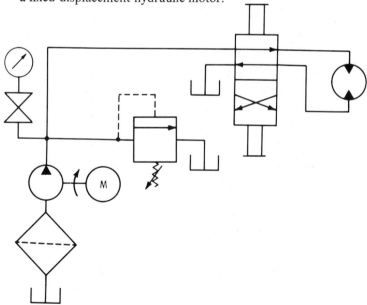

6. A foot-mounted hydraulic motor should be mounted securely to a mounting plate that is rugged enough to withstand the torque that the motor can produce. Use mounting screws or bolts with lock washers to fasten the motor onto the mounting plate. The plate should be flat.

Chapter 17

ANSWERS TO REVIEW QUESTIONS

1. An accumulator is a storage device which is used to store liquid under pressure.

2. A gravity-type accumulator is one in which a mass acts against a ram or piston to put pressure on the liquid.

3. A gravity-type accumulator provides the same pressure for the full stroke of the ram of the accumulator; whereas, in the spring-type accumulator, high pressure is available when the spring is compressed, but the pressure phases out as the spring is completely released. The gravity-type accumulator is larger and delivers more fluid.

4. Advantages of an air- or gas-type accumulator are:
   a. Very compact.
   b. Light in weight.
   c. Low maintenance.
   d. Relatively inexpensive.

5. A piston-type accumulator differs from a hydraulic cylinder in that the piston-type accumulator:
   a. Does not have a piston rod, as does the hydraulic cylinder.
   b. Has a charging valve, but the hydraulic cylinder does not.
   c. Is charged on one side of the piston by air pressure; a hydraulic cylinder does not make use of air pressure.

6. Chief advantage of an accumulator in the hydraulic system of a curing press is that an accumulator can reduce the heat in a system by producing the holding pressure, allowing the pump

either to be stopped or to run under no load for long periods of time.

7. A hydraulic circuit can be set up in which the accumulator can be called upon to deliver liquid during an electric power failure or a pump failure. It should be remembered that only a given quantity of liquid can be delivered under pressure.

8. Usually, a gravity-type accumulator is recommended for a central hydraulic system. A gravity-type accumulator is generally designed to deliver a relatively large amount of fluid.

Chapter 18

ANSWERS TO REVIEW QUESTIONS

1. The lubricator in a pneumatic system should be located to provide lubrication to the controls, cylinders, and air motors.

2. Diagram of simple circuit showing location of speed control to control the piston on the return stroke.

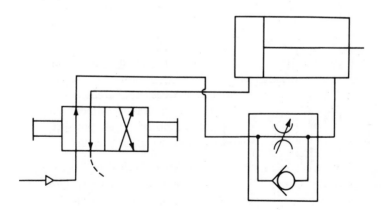

3. Diagram of a circuit using a sequence valve.

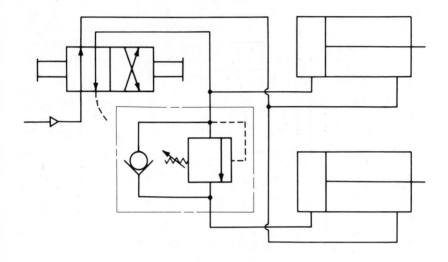

4. Diagram of a circuit using a double-acting cylinder.

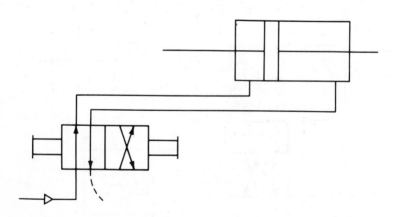

472

5. Diagram of a circuit utilizing two different pressures.

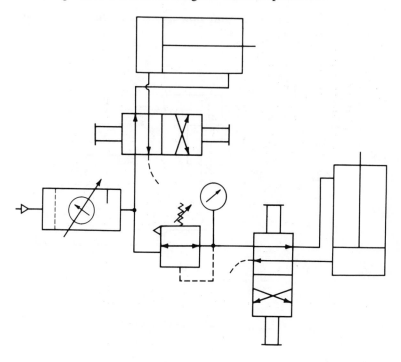

6. Diagram of a circuit utilizing a rotating cylinder.

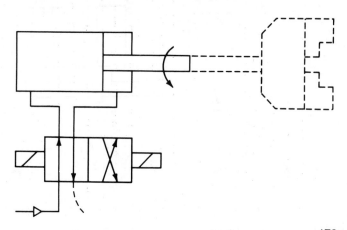

1. Diagram of a circuit showing how a heat exchanger should be connected to a power unit.

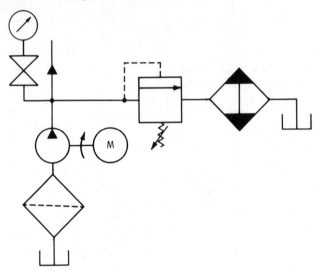

2. Diagram of a circuit using a power unit with two pumps.

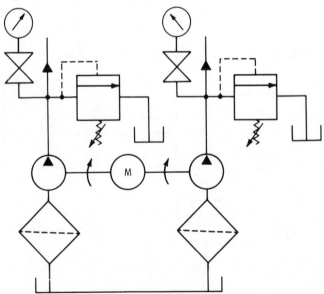

3. Diagram of a circuit in which a hydraulic pilot-operated valve is used to actuate a cylinder.

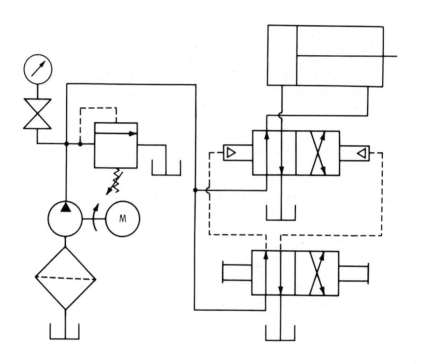

4. Diagram of a circuit utilizing a reducing valve.

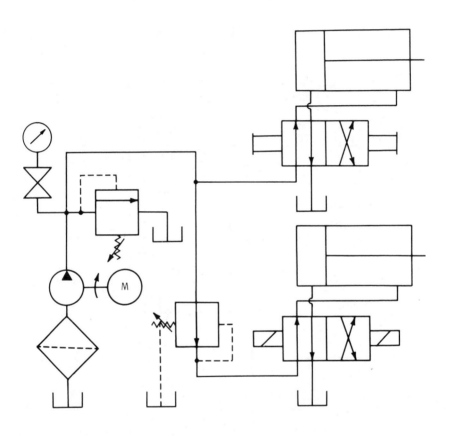

5. Diagram of a circuit utilizing an actuator.

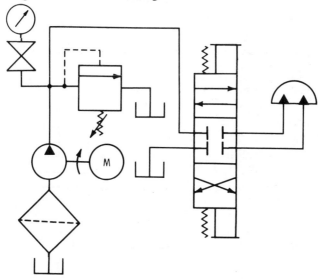

6. Diagram of a circuit in which a hydraulic motor is used in conjunction with a hydraulic cylinder.

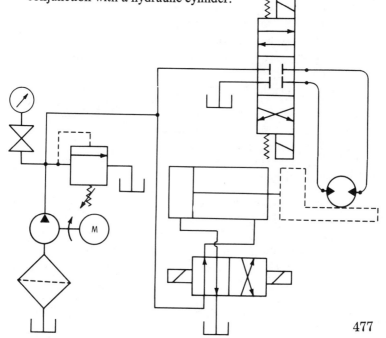

Chapter 20

ANSWERS TO REVIEW QUESTIONS

1. Dirt causes trouble in a hydraulic system by:
   a. Clogging drainage spools of control valves.
   b. Damaging precision parts of pumps.
   c. Scores walls of cylinders.
   d. Causes spools of valves to stick.
   e. Clogs orifices in control calves causing malfunctions.
   f. Clogs intake filters and causes cavitation of pumps.

2. Dirt causes trouble in a pneumatic system by:
   a. Clogs filters, causing reduced flow of air.
   b. Causes internal leaks of controls.
   c. Damages soft seals in cylinders.
   d. Causes spools in valves to stick.
   e. Damages piston rods of cylinders.

3. Effects of heat in a fluid power system are:
   a. Causes excessive internal leakage.
   b. Damages rod seals of cylinders, causing leakage.
   c. Reduces efficiency of pumps.
   d. Causes varnish condition in controls, causing spools to stick.
   e. Causes breakdown of hydraulic oil.

4. Improper oil in a system results in:
   a. Damage to pump by cavitation.
   b. Often causes excessive foaming conditions.
   c. Some oils cause excessive internal leakage.
   d. Sluggishness in control action.

5. In the installation of a hydraulic system, the piping was such that all the drain lines were run into a header that was too small to carry the liquid without building up a back pressure. This caused the functional valves to malfunction. The liquid from the drains should flow back to the reservoir under no back pressure.

6. Air in a hydraulic system can:
   a. Ruin a hydraulic pump.
   b. Cause jerky motion in cylinder's movement.
   c. Cause controls to malfunction.
   d. Cause noise in system.

7. Precautions to observe in installing mufflers in pneumatic systems are:
   a. Use mufflers of adequate size that do not restrict the exhaust of the control to which it is attached.
   b. Make sure the muffler does not become clogged with oil or foreign matter.
   c. Use mufflers that meet the noise abatement requirements.

8. Precautions to observe when a hydraulic system is to be used in an explosive atmosphere are:
   a. Use explosion-proof electric motor on power unit.
   b. Use solenoid-operated control valves that are explosion-proof solenoids.
   c. All electrical controls should be explosion-proof.
   d. For an alternate, place the power unit and electrical controls in a separate room. Use hydraulic controls that are pilot-operated.

9. Choice of hydraulics or pneumatics, as follows:
   a. Hydraulics or a combination of air and hydraulics such as an airdraulic cylinder. Feeds must be accurate and repeatability must be reliable to produce a fine feed on a machine tool.
   b. Hydraulics. These rolls are quite heavy and a large force is required, which can be accomplished with hydraulics.
   c. Pneumatics. Small parts transfer mechanisms usually must work rapidly. Small air cylinders and controls perform well in such an application.
   d. Pneumatics. The rapid expanding of air produces the results needed for a hammer-like blow.
   e. Hydraulics. Molding presses require a large force to produce the desired results.
   f. Pneumatics or hydraulics. Both are used extensively by machine tool builders. Where thru-hole chucking equip-

ment is involved, hydraulics is usually selected, due to the reduction in piston area on thru-hole cylinders.

Chapter 21

ANSWERS TO REVIEW QUESTIONS

1. Five spare parts that should be stocked for a hydraulic system are:
   a. "O" rings.
   b. Piston rod seals for hydraulic cylinders.
   c. Repair kits for control valves which includes all the soft seals.
   d. Spare pump, or at least a pump kit.
   e. Solenoids, or at least solenoid coils for solenoid-operated directional control valve.

2. Stock items for a rotating hydraulic cylinder are:
   a. Piston rod seal.
   b. Piston ring.
   c. Cover gasket.

3. Repair items with a long "shelf life" are: metallic piston rings; some types of piston rod packings; some types of "O" rings; and metal parts, such as valve spools, pump pistons, etc., if properly preserved.

4. Repair items subject to rust should be coated with preservative grease, placed in a sealed plastic bag or wrapped in a coated paper, and stored in a place that is dry.

5. To determine whether the oil in a hydraulic power unit needs changing, contact a reliable oil supplier. He can take samples of oil from a power unit and analyze it in the laboratory to determine the condition of the liquid.

Chapter 22

ANSWERS TO REVIEW QUESTIONS

1. Automatic controls found in most homes are:
   a. A room thermostat. A change in room temperature causes the heating unit either to turn on or to turn off. The governing unit is electrically controlled.
   b. An automatic gas water heater. When the temperature of the water drops to a certain point, the heater turns on; then, when the set temperature is reached, the heater shuts off. The thermostat in the heater causes this action.

2. Automatic controls in most business places are:
   a. The room thermostat, as outlined above.
   b. An automatic door opener. When one steps on a mat in the entrance, the control under the mat signals the door to open. After a certain interval, the door closes automatically. Various types are used.
   c. Many parking lots in big businesses have a similar arrangement. When the front wheels of a car contact a signal, the gate opens and permits the car to pass into the lot; then the gate closes behind the car.

3. Characteristics of a servo system are:
   a. Stability.
   b. Accuracy of response.
   c. Speed of response.

4. Dirt and heat are detrimental to any system. They will cause malfunction of controls.

5. Places where a servo system might be used are:
   a. Steering mechanisms.
   b. Aircraft controls.
   c. Controls on earth moving equipment.
   d. Controls on paper processing equipment.
   e. Controls on machine tools.

INDEX

483

The Audel® Mail Order Bookstore

Here's an opportunity to order the valuable books you may have missed before and to build your own personal, comprehensive library of Audel books. You can choose from an extensive selection of technical guides and reference books. They will provide access to the same sources the experts use, put all the answers at your fingertips, and give you the know-how to complete even the most complicated building or repairing job, in the same professional way.

Each volume:
- **Fully illustrated**
- **Packed with up-to-date facts and figures**
- **Completely indexed for easy reference**

APPLIANCES

REFRIGERATION: HOME AND COMMERCIAL
Covers the whole realm of refrigeration equipment from fractional-horsepower water coolers, through domestic refrigerators to multi-ton commercial installations. 656 pages; 5½ x 8¼; hardbound. **Cat. No. 23286**

AIR CONDITIONING: HOME AND COMMERCIAL
A concise collection of basic information, tables, and charts for those interested in understanding, troubleshooting, and repairing home air conditioners and commercial installations. 464 pages; 5½ x 8¼; hardbound. **Cat. No. 23288**

HOME APPLIANCE SERVICING, 3rd Edition
A practical book for electric & gas servicemen, mechanics & dealers. Covers the principles, servicing, and repairing of home appliances. 592 pages; 5¼ x 8¼; hardbound. **Cat. No 23214**

REFRIGERATION AND AIR CONDITIONING LIBRARY—2 Vols.
Cat. No. 23305

OIL BURNERS, 3rd Edition
Provides complete information on all types of oil burners and associated equipment. Discusses burners—blowers—ignition transformers—electrodes—nozzles—fuel pumps—filters—Controls. Installation and maintenance are stressed. 320 pages; 5½ x 8¼; hardbound. **Cat. No. 23277**

See price list for cost.
All prices are subject to change without notice.
Use the order coupon on the back page of this book.

AUTOMOTIVE

AUTO BODY REPAIR FOR THE DO-IT-YOURSELFER

Shows how to use touch-up paint; repair chips, scratches, and dents; remove and prevent rust; care for glass, doors, locks, lids, and vinyl tops; and clean and repair upholstery. 96 pages; 8½ x 11; softcover. **Cat. No. 23238**

AUTOMOBILE REPAIR GUIDE, 4th Edition

A practical reference for auto mechanics, servicemen, trainees, and owners. Explains theory, construction, and servicing of modern domestic motorcars. 800 pages; 5½ x 8¼; hardbound. **Cat. No. 23291**

CAN-DO TUNE-UP™ SERIES

Each book in this series comes with an audio tape cassette. Together they provide an organized set of instructions that will show you and talk you through the maintenance and tune-up procedures designed for your particular car. All books are softcover.

AMERICAN MOTORS CORPORATION CARS

(The 1964 thru 1974 cars covered include: Matador. Rambler. Gremlin, and AMC Jeep (Willys).). 112 pages; 5½ x 8½; softcover. **Cat. No. 23843**
Cat. No. 23851 Without Cassette

CHRYSLER CORPORATION CARS

(The 1964 thru 1974 cars covered include: Chrysler, Dodge, and Plymouth.) 112 pages; 5½ x 8½; softcover. **Cat. No. 23825**
Cat. No. 23846 Without Cassette

FORD MOTOR COMPANY CARS

(The 1954 thru 1974 cars covered include: Ford, Lincoln, and Mercury.) 112 pages; 5½ x 8½; softcover. **Cat. No. 23827**
Cat. No. 23848 Without Cassette

GENERAL MOTORS CORPORATION CARS

(The 1964 thru 1974 cars covered include: Buick, Cadillac, Chevrolet, Oldsmobile and Pontiac.) 112 pages; 5½ x 8½; softcover. **Cat. No. 23824**
Cat. No. 23845 Without Cassette

PINTO AND VEGA CARS,

1971 thru 1974. 112 pages· 5½ x 8½; softcover. **Cat. No. 23831**
Cat. No. 23849 Without Cassette

TOYOTA AND DATSUN CARS.

1964 thru 1974. 112 pages; 5½ x 8½; softcover. **Cat. No. 23835**
Cat. No. 23850 Without Cassette

VOLKSWAGEN CARS

(The 1964 thru 1974 cars covered include: Beetle. Super Beetle. and Karmann Ghia.) 96 pages; 5½ x 8½; softcover. **Cat. No. 23826**
Cat. No. 23847 Without Cassette

AUTOMOTIVE AIR CONDITIONING

You can easily perform most all service procedures you've been paying for in the past. This book covers the systems built by the major manufacturers, even after-market installations. Contents: introduction—refrigerant—tools—air conditioning circuit—general service procedures—electrical systems—the cooling system—system diagnosis—electrical diagnosis—troubleshooting. 232 pages; 5½ x 8½; softcover. **Cat. No. 23318**

See price list for cost.

All prices are subject to change without notice.

Use the order coupon on the back page of this book.

DIESEL ENGINE MANUAL, 3rd Edition

A practical guide covering the theory, operation, and maintenance of modern diesel engines. Explains diesel principles—valves—timing—fuel pumps—pistons and rings—cylinders—lubrication—cooling system—fuel oil and more. 480 pages; 5½ x 8¼; hardbound. **Cat. No. 23199**

GAS ENGINE MANUAL, 2nd Edition

A completely practical book covering the construction, operation, and repair of all types of modern gas engines. 400 pages; 5½ x 8¼; hardbound. **Cat. No. 23245**

BUILDING AND MAINTENANCE

ANSWERS ON BLUEPRINT READING, 3rd Edition

Covers all types of blueprint reading for mechanics and builders. This book reveals the secret language of blueprints, step-by-step in easy stages. 312 pages; 5½ x 8¼; hardbound. **Cat. No. 23283**

BUILDING MAINTENANCE, 2nd Edition

Covers all the practical aspects of building maintenance. Painting and decorating; plumbing and pipe fitting; carpentry; heating maintenance; custodial practices and more. (A book for building owners, managers and maintenance personnel.) 384 pages; 5½ x 8¼; hardbound. **Cat. No. 23278**

COMPLETE BUILDING CONSTRUCTION

At last—a *one-volume* instruction manual to show you how to construct a frame or brick building from the footings to the ridge. Build your own garage, tool shed, other outbuilding—even your own house or place of business. Building construction tells you how to lay out the building and excavation lines on the lot; how to make concrete forms and pour the footings and foundation; how to make concrete slabs, walks, and driveways; how to lay concrete block, brick and tile; how to build your own fireplace and chimney: It's one of the newest Audel books, clearly written by experts in each field and ready to help you every step of the way. 800 pages; 5½ x 8¼; hardbound. **Cat. No. 23323**

GARDENING & LANDSCAPING

A comprehensive guide for homeowners and for industrial, municipal, and estate groundskeepers. Gives information on proper care of annual and perennial flowers; various house plants; greenhouse design and construction; insect and rodent controls; and more. 384 pages; 5½ x 8¼; hardbound. **Cat. No. 23229**

CARPENTERS & BUILDERS LIBRARY, 4th Edition (4 Vols.)

A practical, illustrated trade assistant on modern construction for carpenters, builders, and all woodworkers. Explains in practical, concise language and illustrations all the principles, advances, and shortcuts based on modern practice. How to calculate various jobs. **Cat. No. 23244**

> Vol. 1—Tools, steel square. saw filing, joinery cabinets. 384 pages; 5½ x 8¼; hardbound. **Cat. No. 23240**
>
> Vol. 2—Mathematics, plans. specifications, estimates 304 pages; 5½ x 8¼; hardbound. **Cat. No. 23241**
>
> Vol. 3—House and roof framing, laying out foundations. 304 pages; 5½ x 8¼; hardbound. **Cat. No. 23242**
>
> Vol. 4—Doors, windows, stairs, millwork, painting. 368 pages; 5½ x 8¼; hardbound. **Cat. No. 23243**

See price list for cost.

All prices are subject to change without notice.

Use the order coupon on the back page of this book.

CARPENTRY AND BUILDING

Answers to the problems encountered In today's building trades. The actual questions asked of an architect by carpenters and builders are answered in this book. 448 pages; 5½ x 8¼; hardbound. **Cat. No. 23142**

WOOD STOVE HANDBOOK

The wood stove handbook shows how wood burned in a modern wood stove offers an immediate, practical, low-cost method of full-time or part-time home heating. The book points out that wood is plentiful, low in cost (sometimes free), and nonpolluting, especially when burned in one of the newer and more efficient stoves. In this book, you will learn about the nature of heat and its control, what happens inside and outside a stove, how to have a safe and efficient chimney, and how to install a modern wood burning stove. You will also learn about the different types of firewood and how to get it, cut it, split it, and store it. 128 pages; 8½ x 11; softcover. **Cat. No. 23319**

HEATING, VENTILATING, AND AIR CONDITIONING LIBRARY (3 Vols.)

This three-volume set covers all types of furnaces, ductwork, air conditioners, heat pumps, radiant heaters, and water heaters, including swimming-pool heating systems. **Cat. No. 23227**

Volume 1

Partial Contents: Heating Fundamentals . . . Insulation Principles . . . Heating Fuels . . . Electric Heating System . . . Furnace Fundamentals . . . Gas-Fired Furnaces . . . Oil-Fired Furnaces . . . Coal-Fired Furnaces . . . Electric Furnaces. **Cat. No. 23248**

Volume 2

Partial Contents: Oil Burners . . . Gas Burners . . . Thermostats and Humidistats . . . Gas and Oil Controls . . . Pipes, Pipe Fitting, and Piping Details . . . Valves and Valve Installations. 560 pages; 5½ x 8¼; hardbound. **Cat. No. 23249**

Volume 3

Partial Contents: Radiant Heating . . . Radiators, Convectors, and Unit Heaters . . . Stoves, Fireplaces, and Chimneys . . . Water Heaters and Other Appliances . . . Central Air Conditioning Systems . . . Humidifiers and Dehumidifiers. 544 pages; 5½ x 8¼; hardbound. **Cat. No. 23250**

HOME MAINTENANCE AND REPAIR: Walls, Ceilings, and Floors

Easy-to-follow instructions for sprucing up and repairing the walls, ceiling, and floors of your home. Covers nail pops, plaster repair, painting, paneling, ceiling and bathroom tile, and sound control. 80 pages; 8½ x 11; softcover. **Cat. No. 23281**

HOME PLUMBING HANDBOOK , 2nd Edition

A complete guide to home plumbing repair and installation. 200 pages; 8½ x 11; softcover. **Cat. No. 23321**

MASONS AND BUILDERS LIBRARY—2 Vols.

A practical, illustrated trade assistant on modern construction for bricklayers, stonemasons, cement workers, plasterers, and tile setters. Explains all the principles, advances, and shortcuts based on modern practice—including how to figure and calculate various jobs. **Cat. No. 23185**

Vol. 1—Concrete, Block, Tile, Terrazzo. 368 pages; 5½ x 8¼; hardbound. **Cat. No. 23182**

Vol. 2—Bricklaying, Plastering, Rock Masonry, Clay Tile. 384 pages; 5½ x 8¼; hardbound. **Cat. No. 23183**

PLUMBERS AND PIPE FITTERS LIBRARY—3 Vols.

A practical, illustrated trade assistant and reference for master plumbers, journeymen and apprentice pipe fitters, gas fitters and helpers, builders, contractors, and engineers. Explains in simple language, illustrations, diagrams, charts, graphs, and pictures, the principles of modern plumbing and pipe-fitting practices. **Cat. No. 23255**

> Vol. 1—Materials, tools, roughing-in. 320 pages; 5½ x 8¼; hardbound. **Cat. No. 23256**

> Vol. 2—Welding, heating, air-conditioning. 384 pages; 5½ x 8¼; hardbound. **Cat. No. 23257**

> Vol. 3—Water supply, drainage, calculations. 272 pages; 5½ x 8¼; hardbound. **Cat. No. 23258**

PLUMBERS HANDBOOK

A pocket manual providing reference material for plumbers and/or pipe fitters. General information sections contain data on cast-iron fittings, copper drainage fittings, plastic pipe, and repair of fixtures. 288 pages; 4 x 6; softcover. **Cat. No. 23339**

QUESTIONS AND ANSWERS FOR PLUMBERS EXAMINATIONS, 2nd Edition

Answers plumbers' questions about types of fixtures to use, size of pipe to install, design of systems, size and location of septic tank systems, and procedures used in installing material. 256 pages; 5½ x 8¼; softcover. **Cat. No. 23285**

TREE CARE MANUAL

The conscientious gardener's guide to healthy, beautiful trees. Covers planting, grafting, fertilizing, pruning, and spraying. Tells how to cope with insects, plant diseases, and environmental damage. 224 pages; 8½ x 11; softcover. **Cat. No. 23280**

UPHOLSTERING

Upholstering is explained for the average householder and apprentice upholsterer. From repairing and regluing of the bare frame, to the final sewing or tacking, for antiques and most modern pieces, this book covers it all. 400 pages; 5½ x 8¼; hardbound. **Cat. No. 23189**

WOOD FURNITURE: Finishing, Refinishing, Repairing

Presents the fundamentals of furniture repair for both veneer and solid wood. Gives complete instructions on refinishing procedures, which includes stripping the old finish, sanding, selecting the finish and using wood fillers. 352 pages; 5½ x 8¼; hardbound. **Cat. No. 23216**

ELECTRICITY/ELECTRONICS

ELECTRICAL LIBRARY

If you are a student of electricity or a practicing electrician, here is a very important and helpful library you should consider owning. You can learn the basics of electricity, study electric motors and wiring diagrams, learn how to interpret the NEC, and prepare for the electrician's examination by using these books. **Cat. No. 23359**

Electric Motors, 3rd Edition. 528 pages; 5½ x 8¼; hardbound. **Cat. No. 23264**

Guide to the 1978 National Electrical Code. 672 pages; 5½ x 8¼; hardbound. **Cat. No. 23308**

House Wiring, 4th Edition. 256 pages; 5½ x 8¼; hardbound. **Cat. No. 23315**

Practical Electricity, 3rd Edition. 496 pages; 5½ x 8¼; hardbound. **Cat. No. 23218**

Questions and Answers for Electricians Examinations, 6th Edition. 288 pages; 5½ x 8¼; hardbound. **Cat. No. 23307**

ELECTRICAL COURSE FOR APPRENTICES AND JOURNEYMEN

A study course for apprentice or journeymen electricians. Covers electrical theory and its applications. 448 pages; 5½ x 8¼; hardbound. **Cat. No. 23209**

See price list for cost.

All prices are subject to change without notice.

Use the order coupon on the back page of this book.

RADIOMANS GUIDE, 4th Edition

Contains the latest information on radio and electronics from the basics through transistors. 480 pages; 5½ x 8¼; hardbound. **Cat. No. 23259**

TELEVISION SERVICE MANUAL, 4th Edition

Provides the practical information necessary for accurate diagnosis and repair of both black-and-white and color television receivers. 512 pages; 5½ x 8¼; hardbound. **Cat. No. 23247**

ENGINEERS/MECHANICS/ MACHINISTS

MACHINISTS LIBRARY, 2nd Edition

Covers modern machine-shop practice. Tells how to set up and operate lathes, screw and milling machines, shapers, drill presses, and all other machine tools. A complete reference library. **Cat. No. 23300**

Vol. 1—Basic Machine Shop. 352 pages; 5½ x 8¼; hardbound. **Cat. No. 23301**

Vol. 2—Machine Shop. 480 pages; 5½ x 8¼; hardbound. **Cat. No. 23302**

Vol. 3—Toolmakers Handy Book. 400 pages; 5½ x 8¼; hardbound. **Cat. No. 23303**

MECHANICAL TRADES POCKET MANUAL

Provides practical reference material for mechanical tradesmen. This handbook covers methods, tools, equipment, procedures, and much more. 256 pages; 4 x 6; softcover. **Cat. No. 23215**

MILLWRIGHTS AND MECHANICS GUIDE, 2nd Edition

Practical information on plant installation, operation, and maintenance for millwrights, mechanics, maintenance men, erectors, riggers, foremen, inspectors, and superintendents. 960 pages; 5½ x 8¼; hardbound. **Cat. No. 23201**

POWER PLANT ENGINEERS GUIDE, 2nd Edition

The complete steam or diesel power-plant engineer's library. 816 pages; 5½ x 8¼; hardbound. **Cat. No. 23220**

QUESTIONS AND ANSWERS FOR ENGINEERS AND FIREMANS EXAMINATIONS, 3RD EDITION

Presents both legitimate and "catch" questions with answers that may appear on examinations for engineers and firemans licenses for stationary, marine, and combustion engines. 496 pages; 5½ x 8¼; hardbound. **Cat. No. 23327**

WELDERS GUIDE, 2nd Edition

This new edition is a practical and concise manual on the theory, practical operation, and maintenance of all welding machines. Fully covers both electric and oxy-gas welding. 928 pages; 5½ x 8¼; hardbound. **Cat. No. 23202**

WELDER/FITTERS GUIDE

Provides basic training and instruction for those wishing to become welder/fitters. Step-by-step learning sequences are presented from learning about basic tools and aids used in weldment assembly, through simple work practices, to actual fabrication of weldments. 160 pages· 8½ x 11; softcover; **Cat. No. 23325**

See price list for cost.

All prices are subject to change without notice.

Use the order coupon on the back page of this book.

FLUID POWER

PNEUMATICS AND HYDRAULICS, 3rd Edition

Fully discusses installation, operation, and maintenance of both HYDRAULIC AND PNEUMATIC (air) devices. 496 pages; 5½ x 8¼; hardbound: **Cat. No. 23237**

PUMPS, 3rd Edition

A detailed book on all types of pumps from the old-fashioned kitchen variety to the most modern types. Covers construction, application, installation, and troubleshooting. 480 pages; 5½ x 8¼; hardbound. **Cat. No. 23292**

HYDRAULICS FOR OFF-THE-ROAD EQUIPMENT

Everything you need to know from basic hydraulics to troubleshooting hydraulic systems on off-the-road equipment. Heavy-equipment operators, farmers, fork-lift owners and operators, mechanics—all need this practical, fully illustrated manual. 272 pages; 5½ x 8¼; hardbound. **Cat. No. 23306**

HOBBY

COMPLETE COURSE IN STAINED GLASS

Written by an outstanding artist in the field of stained glass, this book is dedicated to all who love the beauty of the art. Ten complete lessons describe the required materials, how to obtain them, and explicit directions for making several stained glass projects. 80 pages; 8½ x 11; softbound. **Cat. No. 23287**

BUILD YOUR OWN AUDEL DO-IT-YOURSELF LIBRARY AT HOME!

Use the handy order coupon today to gain the valuable information you need in all the areas that once required a repairman. Save money and have fun while you learn to service your own air conditioner, automobile, and plumbing. Do your own professional carpentry, masonry, and wood furniture refinishing and repair. Build your own security systems. Find out how to repair your TV or Hi-Fi. Learn landscaping, upholstery, electronics and much, much more.

See price list for cost.
All prices are subject to change without notice.
Use the order coupon on the back page of this book.

PRICE LIST

	Price		Price
23142	$10.95	23264	$10.95
23182	9.95	23277	9.95
23183	9.95	23278	9.95
23185	17.95	23280	8.95
23189	9.95	23281	6.95
23199	10.95	23283	9.95
23201	14.95	23284	21.95
23202	14.95	23285	8.95
23209	10.95	23286	12.95
23214	12.95	23287	6.95
23215	8.95	23288	10.95
23216	9.95	23291	14.95
23218	10.95	23292	10.95
23227	32.95	23300	29.95
23229	9.95	23301	10.95
23237	10.95	23302	10.95
23238	6.95	23303	10.95
23240	9.95	23305	21.95
23241	9.95	23306	8.95
23242	9.95	23307	8.95
23243	9.95	23308	12.95
23244	35.95	23315	8.95
23245	9.95	23318	7.95
23247	11.95	23319	7.95
23248	11.95	23321	7.95
23249	11.95	23323	19.95
23250	11.95	23325	7.95
23255	26.95	23327	10.95
23256	9.95	23329	15.95
23257	9.95	23339	8.95
23258	9.95	23359	47.95
23259	11.95		

USE ORDER COUPON ON THE BACK PAGE OF THIS BOOK
ALL PRICES SUBJECT TO CHANGE WITHOUT NOTICE